APPLICATIONS OF GAS CHROMATOGRAPHY

Edited by **Reza Davarnejad**
and **Mahboubeh Jafarkhani**

Applications of Gas Chromatography
Edited by Reza Davarnejad and Mahboubeh Jafarkhani

Published by InTech
Janeza Trdine 9, 51000 Rijeka, Croatia

Publishing Process Manager Daria Nahtigal

Technical Editor Teodora Smiljanic

Cover Designer InTech Design Team

Additional hard copies can be obtained from orders@intechopen.com

Applications of Gas Chromatography,
Edited by Reza Davarnejad and Mahboubeh Jafarkhani
p. cm.
ISBN 978-953-51-0260-1

For EU product safety concerns:
IN TECH d.o.o., Prolaz Marije Krucifikse Kozulić 3, 51000 Rijeka, Croatia,
info@intechopen.com or visit our website at intechopen.com.

Meet the editors

Dr. Reza Davarnejad was born in 1978 (Arak, Iran). He received his bachelor and master in petrochemical and chemical engineering in Iran in 2002 and 2004, respectively. He joined University Sains Malaysia (Penang, Malaysia) and he continued his study in chemical engineering (separation processes). He focused on the supercritical technology, thermodynamics and biotechnology. He was awarded PhD. degree in 2010. He has a lot of publications and research activities. Furthermore, he worked as a process engineer in Shazand Petrochemical Corporation, as a researcher assistant in University Sains Malaysia and as an invited lecturer in several universities. He has joined the Chemical Engineering Department of Arak University as an assistant professor in 2010. He is presently teaching and researching there. Currently, he is as a fixed reviewer and assistant editor in many technical valid journals.

Ms. Mahboubeh Jafarkhani was born in 1985 (Arak, Iran). She received her bachelor and master degree in chemical engineering in Iran in 2007 and 2011, respectively. She researched on tissue engineering and CFD areas and she has several publications. Furthermore, she has an experience in Shazand Oil Refinery.

Contents

Preface

This book presents a critical review of various chromatography techniques for a limited number of processes. Most techniques are illustrated by examples.

The processes described are necessarily limited to those which appear to the authors to have the greatest validity and practical use. Wherever possible, we have included recommendations delineating the best techniques for analyzing each sample. Recommended techniques are often illustrated by detailed examples.

Although the book is intended to serve primarily the practicing engineer, especially the process or chemical engineer, other engineers (such as environmental engineers) and chemists concerned with analyzing techniques may find it useful.

Most new techniques are still empirical in nature, although there are often theoretical bases for the correlation; wherever possible, the theory is outlined to provide the user with the foundation of the proposed chromatography techniques.

Special thanks are due to all respectful authors for their excellent contributions to this book and to Ms. Martina Durovic and Ms. Daria Nahtigal for extensive assistance and support.

Reza Davarnejad and **Mahboubeh Jafarkhani**
Department of Chemical Engineering,
Faculty of Engineering,
Arak University
Iran

1

Gas Chromatography Application in Supercritical Fluid Extraction Process

Reza Davarnejad* and Mostafa Keshavarz Moraveji
Department of Chemical Engineering, Faculty of Engineering, Arak University, Iran

1. Introduction

There are two types of application for gas chromatography (GC) in the supercritical fluid extraction process. Gas chromatography is a type of supercritical extraction apparatuses which can separate a component from a multi-component mixture during supercritical extraction. Therefore, this application can be the alternative to conventional gas chromatography, which needs high temperatures for the evaporation of the feed mixture and for liquid chromatography, where liquid solvents may be replaced. This process results in a different transport velocity along the stationary phase for different molecules. Molecules having weak interaction forces with the stationary phase are transported quickly while others with strong interactions are transported slowly. Beside the interactions with the stationary phase, the solvent power of the mobile phase determines the distribution of the components. Furthermore, supercritical gases have a high solvent power and exert this solvent power at low temperatures.

Another application of GC in supercritical fluid extraction is consideration and analysis of extraction product. The obtained products from various types of supercritical apparatuses (such as phase equilibrium and rate test apparatus) should be analyzed. However, different types of analyzer can be used but, the conventional GC with a suitable column has widely been recommended. Although several columns for detecting a lot of components have been designed and fabricated by some companies but due to lacking of suitable columns for some components or unclear peaks obtained from some columns, an extra process (such as esterification of the fractionated fish oil) before GC analyze is sometimes required. In this application, the samples obtained from the supercritical extraction apparatus are not under pressure or their pressures have broken down by a damper (in online GC).

In this chapter both types of GC application in supercritical fluid extraction with examples will be illustrated.

2. Gas chromatography apparatus

In supercritical fluid chromatography (SFC) the mobile phase is a supercritical gas or a near critical liquid. Compared to gas chromatography (GC), where a gas is under ambient

* Corresponding Author

pressure (for example in the second type of apparatus applied in supercritical process), and liquid chromatography (LC), where a liquid is used as mobile phase, the solvent power of the liquid mobile phase in SFC can be varied by density, e.g., by pressure changes at constant temperature. Solubility increases in general with pressure under supercritical conditions of the mobile phase, temperature sensitive compounds can be processed. The chromatographic separation can be carried out at constant pressure (isobaric operation) or with increasing pressure (pressure programmed). In addition, temperature can be varied. SFC has one more adjustable variable for optimization of elution than GC or LC (Brunner, 1994). A supercritical fluid has properties similar to a gas and also similar to a liquid. While density and solvent power may be compared to those of liquids, transport coefficients are more those of a gas. SFC, because of its mobile phase, can cover an intermediate region between GC and LC, as illustrated in Figure 1 with respect to density and diffusion coefficient. For preparative and production scale operations, SFC has the advantage of easy separation of mobile phase from separated compounds. A disadvantage is that strongly polar and ionic molecules are not dissolved by supercritical gases, which can be advantageously used in SCF (Brunner, 1994).

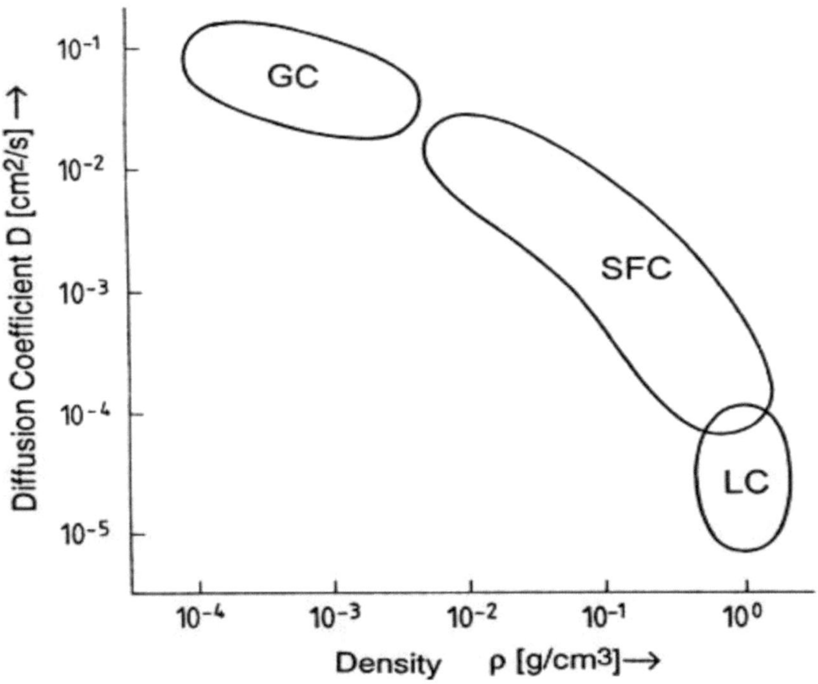

Fig. 1. Areas for the different mobile phases in chromatographic separations with respect to component properties (Schoenmakers and Uunk, 1987).

Most gases which can be used in SFC are non-polar. Therefore, polar substances of a feed mixture can only be eluted by adding a polar modifier. Polar gases like ammonia or sulfur dioxide are reactive compounds under pressure the equipment must be able to withstand corrosive conditions. On the other hand, carbon dioxide is easy to handle and safe. Polar modifiers, which are easier to handle than ammonia or sulfur dioxide may instead be applied. To make effective use of the possibilities of SFC, allowable pressures should be high.

Composition of the mobile phase can substantially influence separation in SFC (Brunner, 1994). Retention times of substances may be very much different due to polarity or other physico-chemical properties of the components of the mobile phase. Pickel (1986) investigated large differences in the separation of aromatic hydrocarbons with CO_2, N_2O, C_3H_8 and C_3H_6.

Gases applied in SFC are mostly non-polar. The polarity of carbon dioxide at low densities is comparable to that of n-hexane and at higher densities to that of methylene chloride. Nitrous oxide and the alkanes butane or pentane behave similar. Polar substances are eluted only after long retention times and in broad peaks even not at all. In these cases, a polar modifier, added to the gaseous mobile phase, introduces the necessary polarity to the mobile phase. The modifier then determines the elution sequence, which can be changed by the amount and the type of modifiers (Brunner, 1994).

With increasing content of a modifier in the mobile phase, retention times become shorter. For polycyclic aromatic compounds, Leyendecker *et al.* (1986) investigated the influence of 1.4-dioxane as modifier on n-pentane as mobile phase.

Temperature and pressure can be employed in supercritical chromatography as parameters for influencing separation characteristics. Temperature directly determines vapor-pressure of the feed components and density of the mobile phase and, indirectly, adsorption equilibrium. With higher temperatures, vapor-pressures of the feed components increase exponentially. Density decreases proportionally to temperature if conditions are far from critical, but in the region of the critical point of the mobile phase, which is the main area of application of SFC, density varies dramatically with temperature. The solvent power of the mobile phase, which increases with density, is therefore changed substantially in this region. The influence on chromatographic separation depends on the relative importance of these two effects, if other conditions remain unchanged (Brunner, 1994).

Pressure mainly influences density of the mobile phase. With increasing pressure, the influence of temperature is diminishing, since density varies less with temperature at higher pressures (Brunner, 1994). SFC allows the variation of temperature and pressure for optimizing separation conditions as well as during the separation process itself. Such an operational mode is called pressure and temperature programming. Temperature programming is well known from gas chromatography, but is less common in SFC, since pressure programming can be very effective. Pressure and temperature programming may be combined to density programming (Brunner, 1994).

In preparative chromatography, conditions are kept constant during separation, since feed mixtures of several injections may be on their way at the same time in the column. The elution of substances of different molecular weight in isobaric SFC separations is better than in isothermal GC, since vapor pressure is not so important in SFC. Compared to LC, the tendency of peak broadening is lower in SFC, since diffusion coefficients are far higher (Brunner, 1994).

Flow rate of the mobile phase is a further important parameter which affects the number of theoretical stages in chromatographic separations. Due to the low viscosity of near critical mobile phases, flow rate in SFC can be high, and number of theoretical stages remains nearly constant over a wide range of flow rate. A more detailed discussion of chromatographic fundamentals and especially analytical applications of SFC can be found in the abundant literature on analytical SFC (Gere et al., 1982; Lee and Markides, 1990; Smith, 1988; Wenclawiak, 1992; White, 1988).

The apparatus (as shown in Figure 2) consists of the separation column as central part in a temperature controlled environment (1), the reservoir for the mobile phase (2), a unit for establishing, maintaining and controlling pressure (3), an optimal unit for adding a modifier (4), the injection part for introducing the feed mixture (5), a measuring device (detector) for determining concentration of the eluted substances (6), a sample collection unit (7), a unit for processing the mobile phase (8) and another one for processing data and controlling the total apparatus (9) (Brunner, 1994).

The flow of the supercritical gas under pressure is maintained by long-stroke piston-pumps, reciprocating piston-pumps or membrane-pumps which deliver the mobile phase in liquefied form. The fluid is then heated to supercritical conditions before entering the column. Pressure and flow rate must be kept as constant as possible in order to maintain constant conditions for separation and to achieve a stable base line in the chromatogram. Oscillating pumps therefore can have three heads which deliver at different times or a pulsation dampener in order to minimize pulsation (Brunner, 1994).

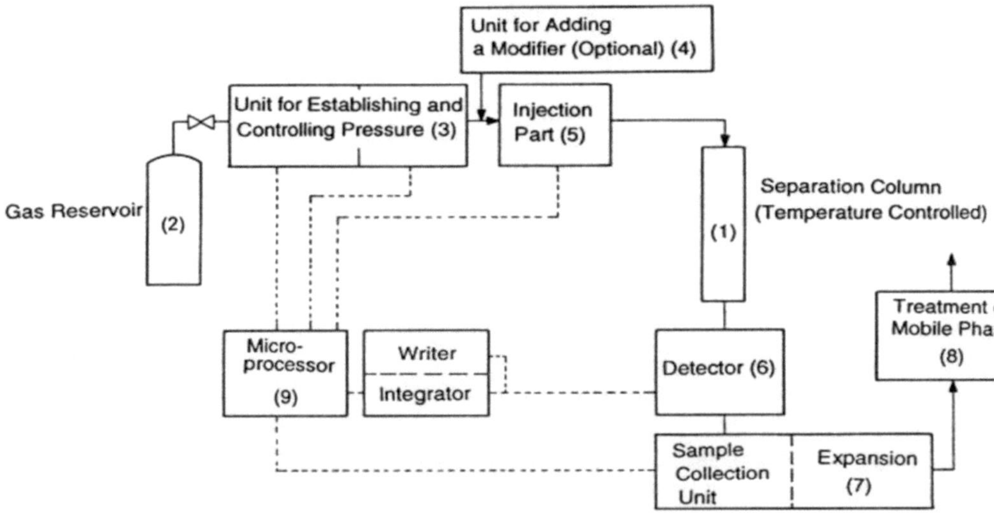

Fig. 2. Flow scheme of apparatus for SFC (Brunner, 1994).

2.1 Columns

Columns for chromatographic separation with supercritical are chosen, like other chromatographic columns, according to the needs of the separation. For analytical purposes the choice is between packed and capillary column. Capillary columns are used with a length between 10 m and 25 m. Pressure drop is low compared to liquid mobile phases. Therefore, capillary columns with inner diameters of 50 to 100 μm can be used and a high number of theoretical stages verified. Separations with capillary columns can be nearly as effective as in gas chromatography. While in early applications steel capillaries had been used in SFC, since 1980, fused silica capillary columns have replaced the steel capillaries. Stationary phases mostly stem from polysiloxanes and polyglycols. Frequently used stationary phases have been listed in the literature (Brunner, 1994).

Under conditions of SFC, the compounds of the stationary phases may be slightly soluble in the mobile phase and are therefore fixed by linking them by chemical reactions. Numerous packed columns are available, many from HPLC applications. Normal phase chromatography (polar stationary phase, non-polar mobile phase) and reverse phase chromatography (non-polar stationary phase, polar mobile phase) are applied, but are not as important in SFC as in HPLC, since a polar mobile phase in SFC involves a polar modifier. Most separations in SFC are carried out with unmodified silica gel or chemically modified silica gel as stationary phase (Brunner, 1994). For packed columns particles are available with diameters in the range of 3 to 100 μm. For analytical purposes particles in the range of 3 to 5 μm have a high separation power in a packing and enable a high linear velocity of the mobile phase leading to short retention times. For preparative purposes particles in the range of 20 to 100 μm are used (Brunner, 1994).

Special filling techniques are necessary to ensure a homogeneous packing. Saito and Yamauchi (Saito *et al.*, 1988; Saito and Yamauchi, 1988; Saito et al., 1989) and Yamauchi and Saito (Yamauchi et al., 1988; Yamauchi and Saito, 1990) applied columns of 7 to 20 mm diameter, Perrut (1982, 1983, 1984) a column of 60 mm inner diameter and 600 mm length with particles of 10-25 μm, Alkio *et al.* (1988) a 900 mm long column with 40-36 μm diameter particles.

The length of the column is dependent on the allowable pressure drop. Pressure drop usually is in the range of 1 to 4 MPa for 250 mm. In this range for the pressure drop capillary factors are nearly independent of pressure drop as demonstrated by Schoenmakers *et al.* (1986). To avoid unacceptable pressure drop, Saito *et al.* applied a recycling technique (Saito *et al.*, 1988; Saito and Yamauchi, 1988; Saito *et al.*, 1989). A cycle pump transports the eluted substances several times to the beginning of the column. Thus, the separation power of the column can be enhanced, without increasing pressure drop. Peak broadening occurs due to the cyclic operations.

2.2 Detectors

Detection of a substance is necessary in analytical and preparative chromatography. In general, the same detectors are used as in gas and liquid chromatography. Selection of a detector depends on the quantity of substance available and the chemical nature the compound. A flame ionization detector (FID) detects substances down to nanogram quantities. Between two electrodes a voltage of 300 V and a hydrogen flame are maintained. If a substance with at least one carbon-hydrogen bonding is eluted from the column to the detector, it is burned and ions are formed, which leads to a current between the electrodes. The current is amplified and processed as a signal for the concentration of the substance. Nearly all substances can be detected. Response factors mainly differ according to number of carbon atoms, therefore calibration is easy (Brunner, 1994).

The ultraviolet spectroscopy detector (UV) is a nondestructive detector, which can be applied at column pressure. It is widely used, but is limited to substances with chromophoric groups. Saturated hydrocarbons, fatty acids and glycerides may be difficult to detect quantitatively. These substances may be detected with a refractive increment detector (RID), where the variation of refractive index of the mobile phase caused by dissolved substances is applied for detection. Other detectors are the fluorescence detector and the light-scattering detector (Brunner, 1994).

In a light-scattering detector the mobile phase is intensively mixed with an inert gas and heated while flowing downward a tube (Upnmoor and Brunner, 1989; Upnmoor and Brunner, 1992). The inert gas and the temperature increase reduce solvent capacity of the mobile phase. The eluted substances precipitate and are carried out droplets or particles into the detection chamber. Into this chamber a tungsten lamp delivers visible light, which is dispersed by droplets or particles. The dispersed light is detected by a photomultiplier under an angle of 60°. The signal is proportional to the mass of light-scattering particles. Therefore, the light-scattering detector acts as mass detector and its signal is independent on chromatographic groups. It can be applied for detection of chromatographic and non-chromatographic substances in a mixture, as for example, in fatty acids and glycerides (Brunner, 1994).

2.3 Expansion of mobile phase and sample collecting system

In analytical SFC, the mobile phase is either expanded after or before detection. Downstream to a detector, which is operated under column pressure, expansion can be achieved by normal expansion valves. They can act as back pressure regulators may be controlled by a central unit. At interesting alternative to an expansion valve was designed by Saito and Yamauchi, who use time-controlled opening and closing of an unrestricted tube for expansion. This has the advantage that blocking of the tube by precipitating substances is avoided. Another expansion technique was adapted from GC: A glass capillary is formed into long, thin capillary, as so-called restrictor. Problems with blocking and difficulties with reproducible manufacturing of the restrictors are disadvantages of this solution. In analytical SFC expansion techniques are determined by detection needs. The amount of substances is small and can easily be handled. The quantity of the mobile phase is not small and it must be recycled. To avoid backmixing, the recycled mobile phase must be totally free from any dissolved substances. In most cases they will be in the range of 0.1 or even 0.01% (Brunner, 1994). Then, separation methods for the dissolved substances from the mobile phase become important. Figure 3 shows a chromatographic system proposed by Perrut (1982, 1983, 1984). After elution and detection, the mobile phase together with the dissolved substance is heated and expanded. By these means the solvent power of the mobile phase is reduced and the substance precipitates; it is collected in one of several collecting vessels, one for each substance. The substances are removed after sufficient quantities of each of the substances have accumulated after several injections. Before the expanded mobile phase can be recycled by a cycle pump, it is passed through an adsorbing bed, where remaining quantities of the dissolved substances and other un wanted substances (as, for example, water) are removed. As in any solvent cycle, make up gas must be added, and a small part of the solvent must be removed for disposal or for special cleaning (Brunner, 1994).

In preparative SFC so far mostly extracts from plants like lemon peel oil, tocopherols from wheat germ or ubichinones have been treated. Unsaturated fatty acids from fish oil, mostly processed as esters, is a subject investigated heavily in recent years (Davarnejad *et al.*, 2008).

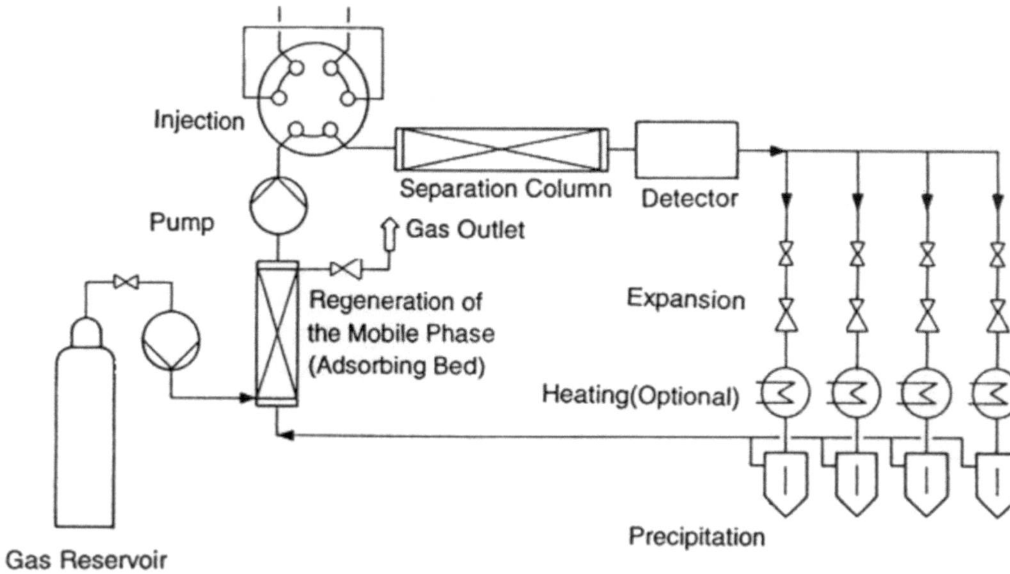

Fig. 3. Flow scheme of a preparative SFC with recycle of the mobile phase (Perrut, 1982, 1983, 1984).

More specialized applications deal with polymers or the fractionation of coal tar. Berger and Perrut (1988) have reviewed the applications of preparative SFC.

2.4 Injection techniques

Injection of the mixture to be separated is accomplished for analytical purposes by sample loops which may be filled at ambient pressure and are injected into the flow of the mobile phase by switching a multiposition valve in the appropriate position. Such valves can be manufactured as linear moving or rotating valve, as shown in Figure 4.

For preparative separations the feed is pumped by metering pumps into the flow of the mobile phase. Intensive mixing can be achieved in line by static mixers. Other possibilities comprise a column, where the mixture is placed under ambient pressure and is then eluted by the mobile phase and transported to the separation column, or a combination with a gas extraction unit. The extract of the gas extraction process can be directly passed through the chromatographic column. The separated substances can be collected. The extract from an extraction unit is diluted with respect to the interesting compounds. It can be collected on a column and after some time transported to the chromatographic separation (Brunner, 1994). This operational mode is illustrated in Figure 5.

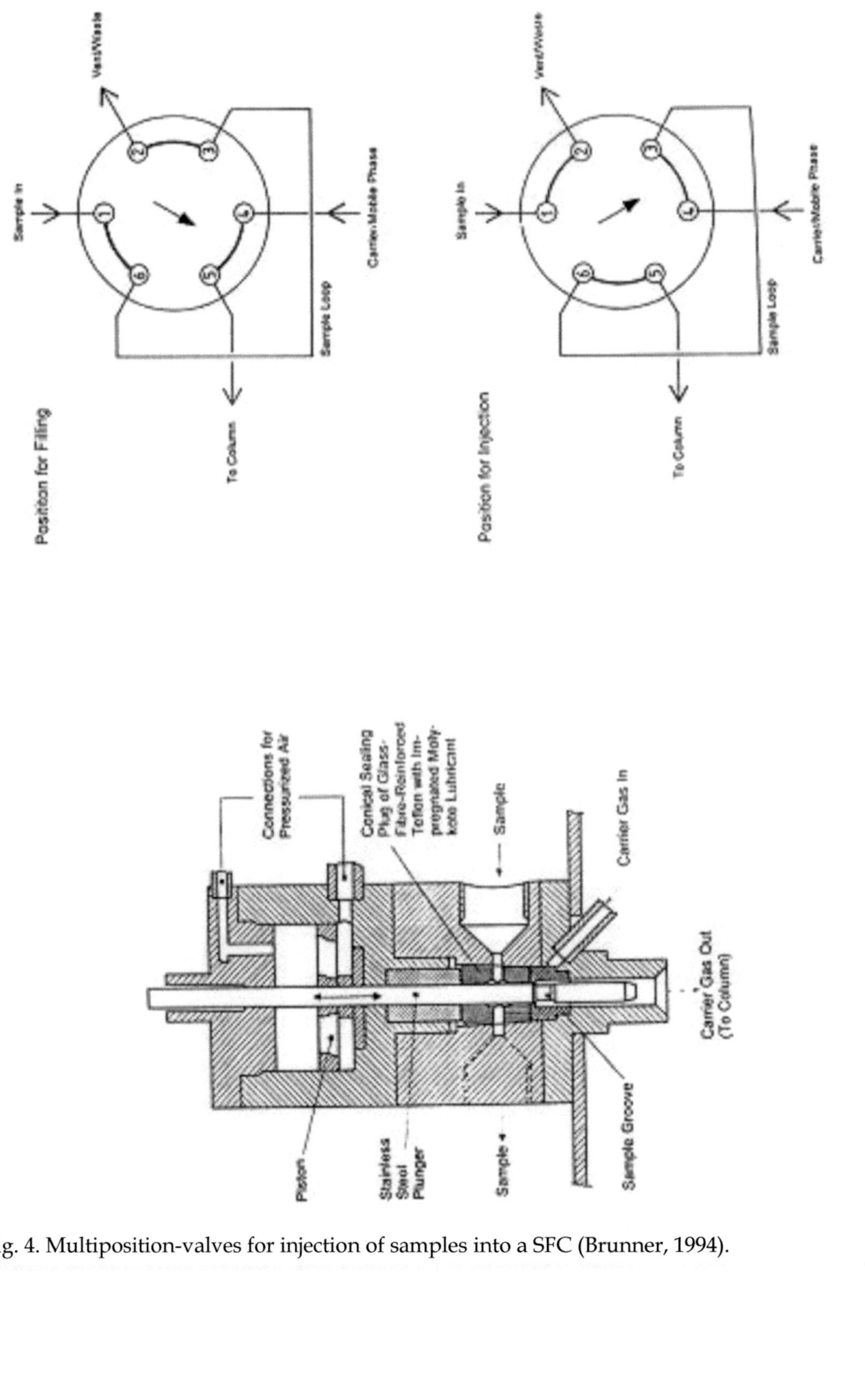

Fig. 4. Multiposition-valves for injection of samples into a SFC (Brunner, 1994).

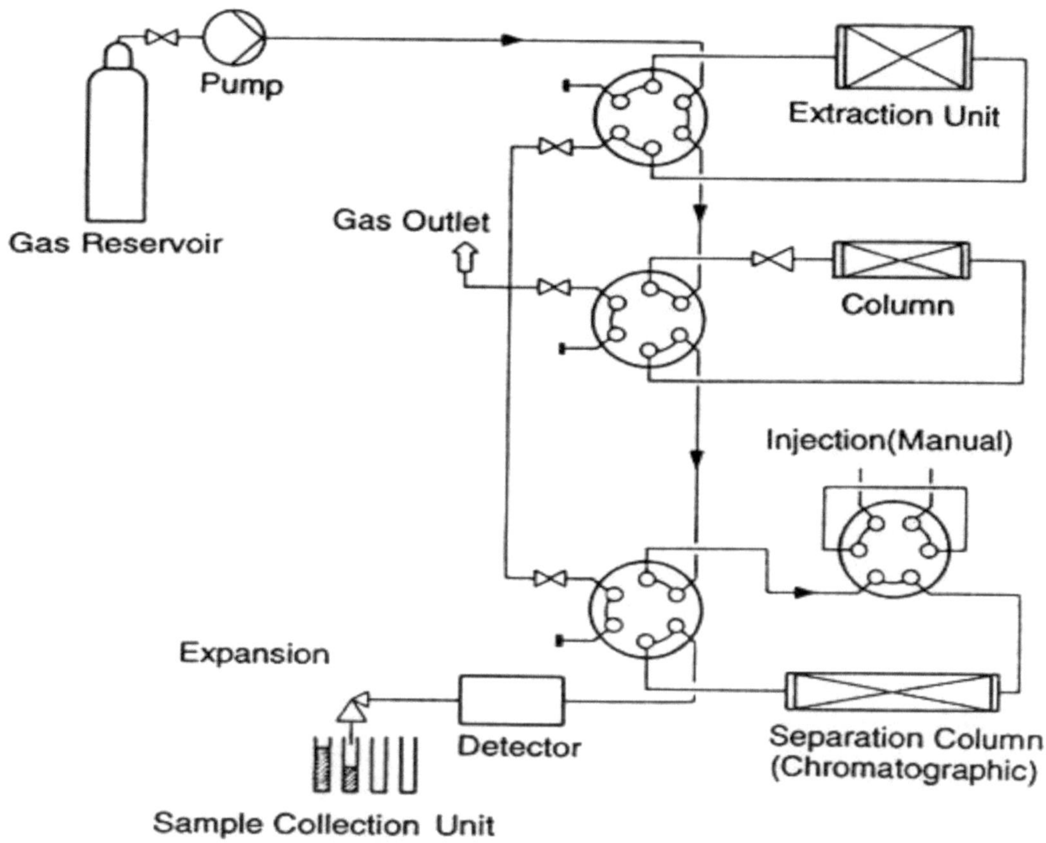

Fig. 5. Coupling of SFE with SFC. Concentration of compounds in a collecting column (Yamauchi and Saito, 1990).

According to this type of apparatus, some examples in details have been shown by Brunner (1994).

3. General gas chromatography apparatus

The obtained products from various types of supercritical apparatuses (such as phase equilibrium and rate test apparatus) should be analyzed. However, different types of analyzer can be used, but the conventional GC with a suitable column has widely been recommended. Although several columns for detecting a lot of components have been designed and fabricated by some companies, but due to lacking of suitable columns for some components or unclear peaks obtained from some columns, an extra process (such as esterification of the fractionated fish oil) before GC analyze is required. In this application, the samples obtained from the supercritical extraction apparatus are not under pressure or their pressures have broken down by a damper (in the online GC).

Since this type of apparatus has been explained in detail in the other chapters, therefore an example from its application is illustrated in this section.

3.1 Triacylglycerols analysis

3.1.1 Introduction

Most of the fatty acids of palm oil are present as triacylglycerols (TAGs). The different placements of fatty acids and fatty acid types on the glycerol molecule produce a number of different TAGs. About 7 to 10 percent of saturated TAGs are predominantly tripalmitin (Karleskind and Wolff, 1996) and the fully unsaturated triglycerides constitute 6 to 12 percent (Karleskind and Wolff, 1996; Kifli, 1981). The TAGs in palm oil are partially defined most as of the physical characteristics of the palm oil such as the melting point and crystallization behavior (Sambanthamurthi *et al.*, 2000). Detailed information about Malaysian tenera palm oil TAGs have been given in various references (Sambanthamurthi *et al.*, 2000; Kifli, 1981; Sow, 1979). Fatouh *et al.* (2007) studied the supercritical extraction of TAGs from buffalo butter oil using carbon dioxide solvent. They concluded that increasing the pressure and temperature of the extraction led to increasing the solvating power of the supercritical carbon dioxide. In these studies, the TAGs were extracted during the early stage of the fractionation thereby creating low-melting fractions. Conversely, TAGs were concentrated in the fractions (i.e. high-melting fractions) obtained towards the end of the process.

According to the literature, mole fraction solubility data of pure triacylglycerols in CO_2 were reported at temperatures of 40, 60 and 80 °C in the range of 10^{-10} to 10^{-2} (Soares *et al.*, 2007). These data depended on the type of triacylglycerols and operating pressure. That means high pressure had a good effect on solubility of triacylglycerols in CO_2. Furthermore, tricaprylin had the higher solubility in CO_2 (around 10^{-2} at high pressures) than the rest (Jensen and Mollerup, 1997; Bamberger *et al.*, 1988; Weber *et al.*, 1999).

In this research, phase equilibrium of TAGs from crude palm oil in sub and slightly supercritical CO_2 is studied. For this purpose, the samples obtained from the phase equilibrium supercritical fluid extraction apparatus are carefully analyzed by a HPLC in terms of TAGs.

3.1.2 Experiment

3.1.2.1 Materials and methods

Crude palm oil and CO_2 (99.99%) respectively as feed and solvent were purchased from United Oil Palm Industries, Nibong Tebal, Malaysia and Mox Sdn. Bhd. 1,3-dipalmitoyl-2-oleoyl-glycerol (99%) (POP), 1,2-dioleoyl-3-palmitoyl-rac-glycerol (99%) (POO) and 1,2-dioleoyl-3-stearoyl-rac-glycerol (99%) (SOO) as standards were purchased from Sigma-Aldrich. Acetone (99.8%) and acetonitrile (99.99%) as solvent and mobile phase were obtained from J.T.Baker and Fisher Scientific, respectively.

The phase equilibrium supercritical fluid extraction apparatus and calculations procedure have been shown and explained in detail in the references (Davarnejad, 2010; Davarnejad *et al.*, 2010; Davarnejad *et al.*, 2009).

The operating conditions were set at 10.8, 7.0, 2.7 and 1.7 MPa for temperature of 80 °C, 11.1, 7.6, 6.1 and 1.1 MPa for temperature of 100 °C and 7.4, 5.4, 3.3 and 0.6 MPa for temperature of 120 °C.

For TAGs analysis, the following stages are carried out step by step using a HPLC (brand: Shimadzu, Japan; model: 10 Series) which is equipped with a capillary column (Aglient Lichrosphere RP-18250×4 mn) and oven temperature is also set at 50 °C.

1. The standard solutions of POO, POP and SOO with concentrations of 10, 25 and 50 ppm are prepared by diluting these chemicals with acetone separately. These solutions are prepared by diluting these chemicals which are initially prepared at a concentration of 200 ppm.
2. A mobile phase containing 75% acetone and 25% acetonitrile (v/v) is prepared.
3. The chromatography interface, vacuum degasser, pump (LC-8A pump with maximum flow rate of 150 cm^3/min) and refractive index (RI) detector are switched on respectively.
4. The computer, printer and GC are switched on.

The HPLC diagram obtained from each sample is shown as following and compared with the diagrams obtained from the standard materials. Then, by applying the standard equation:

Samples concentration (ppm)=sample area/standard area ×standard concentration

Samples concentration in terms of each substance is calculated.

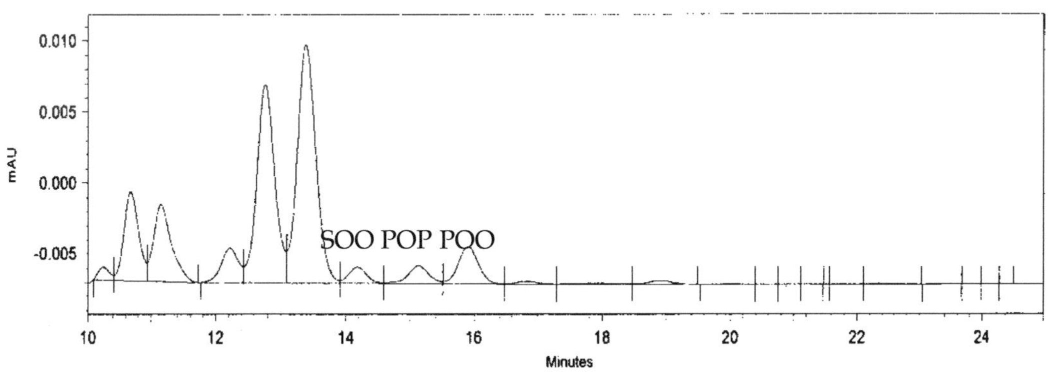

Fig. 6. HPLC chromatogram of crude palm oil TAGs for vapor phase at pressure of 5.4 MPa and temperature of 120 °C.

3.1.3 Results and discussion

Two-phase equilibrium data based on different TAGs are illustrated in Tables 1-3 as following:

Pressure (MPa)	Liquid phase, CO_2 mole fraction	Vapor phase, CO_2 mole fraction
T=80 °C		
10.8	0.9	1
7.0	0.9	1
2.7	1	1
1.7	1	1
T=100 °C		
11.1	1	1
7.6	1	1
6.1	1	1
1.1	1	1
T=120 °C		
7.4	0.9	1
5.4	0.9	1
3.3	1	1
0.6	1	1

Table 1. Two-phase equilibrium calculated data based on CO_2 at 80,100 and 120 °C in order to POP analysis

Pressure (MPa)	Liquid phase, CO_2 mole fraction	Vapor phase, CO_2 mole fraction
T=80 °C		
10.8	0.9	1
7.0	0.9	1
2.7	1	1
1.7	1	1
T=100 °C		
11.1	1	1
7.6	1	1
6.1	1	1
1.1	1	1
T=120 °C		
7.4	0.9	1
5.4	0.9	1
3.3	1	1
0.6	1	1

Table 2. Two-phase equilibrium calculated data based on CO_2 at 80,100 and 120 °C in order to POO analysis

Pressure (MPa)	Liquid phase, CO_2 mole fraction	Vapor phase, CO_2 mole fraction
T=80 °C		
10.8	0.9	1
7.0	0.9	1
2.7	1	1
1.7	1	1
T=100 °C		
11.1	0.9	1
7.6	1	1
6.1	1	1
1.1	1	1
T=120 °C		
7.4	0.9	1
5.4	0.9	1
3.3	1	1
0.6	1	1

Table 3. Two-phase equilibrium calculated data based on CO_2 at 80,100 and 120 °C in order to SOO analysis

The mole fractions of CO_2 in the equilibrium supercritical extraction of the POP substance for the liquid and vapor phases are shown in Table 1. The POP substance mole fractions in the vapor phase increased with increasing the pressure at 80 °C. This trend was also observed in the liquid phase. A regular trend was not observed in the vapor and liquid phases at 100 °C and 120 °C. It seems that high temperatures (such as 100 °C and 120 °C) to cause this irregularity however, the maximum solubility was carried out at 120 °C which is reasonable because high temperature increases TAGs solubilities in CO_2 (Soares *et al.*, 2007). The maximum solubility of POP substance in CO_2 was observed at 80 °C and 10.8 MPa, at 100 °C and 7.6 MPa as well as at 120 °C and 5.4 MPa, respectively. The optimum conditions around these operating regions for the maximum solubility of the POP substance were at 120 °C and 5.4 MPa.

The mole fractions of CO_2 in the equilibrium supercritical extraction of the POO substance for the liquid and vapor phases are shown in Table 2. The POO substance mole fractions in the vapor phase increased with increasing the pressure at 80 °C. This trend was also observed in the liquid phase. A regular trend was not observed in the vapor and liquid phases at 100 °C and 120 °C. Since POO is from TAGs group, this subject is reasonable as it was legitimized for POP. The maximum solubility of the POO substance was observed at 7.6 MPa and 100 °C as well as at 5.4 MPa and 120 °C, respectively. The optimum conditions around these operating regions for the maximum solubility of the POO substance was observed at 120 °C and 5.4 MPa.

The mole fractions of CO_2 in the equilibrium supercritical extraction of the SOO substance for the liquid and vapor phases are shown in Table 3. The SOO substance mole fractions in the vapor phase increased with increasing the pressure at 80 °C; this trend was also observed in the liquid phase. A regular trend was not observed in the vapor and liquid phases at 100 °C and 120 °C. Since SOO also is from TAGs group, this subject is reasonable as it was legitimized for POP and POO. The maximum solubility of the SOO substance was observed at 7.6 MPa and 100 °C as well as at 5.4 MPa and 120 °C. The optimum conditions around these operating regions for the maximum solubility of the SOO substance was observed at 120 °C and 5.4 MPa.

The experimental results showed that although highest extraction using CO_2 was observed at 120 °C and 5.4 MPa for all of the TAGs substances but, maximum solubilities at the mentioned operating conditions were obtained for POP (y=1), POO (y=1) and SOO (y=1) respectively. The obtained result was confirmed by the overall conclusion which stated that the fats and oils composed of shorter chain fatty acids are more soluble than those with longer chain fatty acids (Fatouh *et al.*, 2007). Soares *et al.* (2007) also illustrated that high temperature had a desired effect on TAGs solubility in CO_2 as this output was obtained from the current research as well where this statement supported the results above.

Tan *et al.* (2008) also confirmed that high temperature had a positive effect on TAGs extraction using CO_2. They observed that monounsaturated fatty acids (such as POP) had a maximum yield of extraction in supercritical CO_2 as this output was obtained from current research. That means POP had a higher solubility in CO_2 in comparison with POO and SOO. Although they proposed high pressures for supercritical extraction of TAGs but they conducted their experiments at operating pressures more than 10 MPa and they studied yield of extraction.

3.1.4 Conclusions

In the mutual solubility study of TAGs from crude palm oil which is related to the experiment, it was shown that by using the supercritical fluid extraction process, the highest mole fraction percentage of TAGs solubility was obtained at approximately 2.2% (at 5.4 MPa and 120 °C). According to our calculations, the data slightly varied in seventh or eighth decimal points for the reported data due to the chromatography influence. In order to the calculations procedure, the significant part of these calculated data (solubility data) is prepared from the pressure increments. Furthermore, high temperature increased TAGs solubility in CO_2.

4. References

Alkio, M. Harvala T. Komppa, V. (1988). Preparative scale supercritical fluid chromatography. In: Perrut, M. (ed.) *Proceedings of 2nd International Symposium on Supercritical Fluids.*

Bamberger, T. Erickson, J.C. Cooney, C.L. Kumar, S.K. (1988). Measurement and model prediction of solubilities of pure fatty acids, pure triglycerides, and mixtures of triglycerides in supercritical carbon dioxide, *J. Chemical & Engineering Data*, Vol. 33, No. 3, pp. 327–333.

Berger, C. Perrut, M. (1988). Purification de molecules d'interet biologique par chromatografie preparative avec eluant supercritique, *Technoscope Biofutur*, Vol. 25, pp. 3-8.

Brunner, G. (1994). Gas Extraction: an Introduction to Fundamentals of Supercritical Fluids and the Application to Separation Processes, Springer, 0-387-91477-3, New York.

Davarnejad, R. Kassim, K.M. Ahmad, Z. Sata, S.A. (2009). Solubility of β-carotene from crude palm oil in high temperature and high pressure carbon dioxide, *J. Chemical & Engineering Data*, Vol. 54, No. 8, pp. 2200-2207.

Davarnejad, R. Ahmad, Z. Sata, S.A. Moraveji, M.K. Ahmadloo, F. (2010). Mutual solubility study in supercritical fluid extraction of tocopherols from crude palm oil using CO_2 solvent, *International J. Molecular Sciences*, Vol. 11, No. 10, pp. 3649-3659.

Davarnejad, R. Kassim, K.M., Ahmad, Z. Sata S.A. (2008). Extraction of fish oil by fractionation through the supercritical carbon dioxide, *J. Chemical & Engineering Data*, Vol. 53, No. 9, pp. 2128-2132.

Davarnejad, R. (2010). Phase equilibrium study of β-carotene, tocopherols and triacylglycerols in supercritical fluid extraction process from crude palm oil using carbon dioxide as a solvent, PhD. thesis, Universiti Sains Malaysia, Malaysia.

Fatouh, A.E. Mahran, A. El-Ghandour, M.A. Singh, R.K. (2007). Fractionation of buffalo butter oil by supercritical carbon dioxide. *LWT*, Vol. 40, No. 10, pp. 1687-1693.

Gere, D. Board, R. Mc-Manigill, D. (1982). Parameters of Supercritical Fluid Chromatography Using HPLC Columns, Hewlett-Packard, 43-5953-1647, Avondale, Pennsylvania.

Jensen, C.B. Mollerup, J. (1997). Phase equilibria of carbon dioxide and tricaprylin, *J. Supercritical Fluids*, Vol. 10, No. 2, pp. 87–93.

Karleskind, A. Wolff, JP. (1996). In: volume 1: Properties, Production and Applications, Intercept Limited, Andover, UK.

Kifli, H. (1981). Studies on palm oil with special reference to interesterification. PhD. Thesis, University of St Andrews, Scotland.

Lee, M.L. Markides, K.E. (eds) (1990). Analytical supercritical fluid chromatography and extraction, *Proceedings of Chromatography*, Provo, Utah.

Leyendecker, D. Leyendecker, D. Schmitz, F.P. Klesper, E. (1986). Chromatographic behavior of various eluents and eluent mixtures in the liquid and in the supercritical state. *J. Chromatography A*, Vol. 371, pp. 93-107.

Perrut, M. (1982). Demande de brevet francais, No. 82 09 649.

Perrut, M. (1983). Brevet European, No. 00 99 765.

Perrut, M. (1984). US Patent, No. 44 78 720.

Pickel, K.H. (1986). Chromatografische Untersuchungen mit Hochkompressiblen Mobilen Phasen, Dissertation, Universität Erlangen-Nürnberg.

Saito, M. Yamauchi, Y. Hondo, T. Senda M. (1988). Laboratory scale preparative supercritical chromatography in recycle operation: instrumentation and applications. In: Perrut, M. (ed.), *Proceedings of 2nd International Symposium on Supercritical Fluids.*

Saito, M. Yamauchi, Y. (1988). Recycle chromatography with supercritical carbon dioxide as mobile phase, *J. High Resolution Chromatography*, Vol. 11, No. 10, pp. 741-743.

Saito, M. Yamauchi, Y. Inomata, K. Kottkamp, W. (1989). Enrichment of tocopherols in wheat germ by directly coupled supercritical fluid extraction with semi-preparative supercritical chromatography, *J. Chromatographic Science*, Vol. 27, No. 2, pp. 79-85.

Saito, M. Yamauchi, Y. (1990). Isolation of tocopherols from wheat germ oil by recycle semi-preparative supercritical fluid chromatography, *J. Chromatography A*, Vol. 505, No. 1, pp. 257-271.

Schoenmakers, P.J. Verhoeven, F.C.C.J.G. (1986). Effect of pressure on retention in supercritical fluid chromatography with packed columns, *J. Chromatography A*, Vol. 325, pp. 315-328.

Sambanthamurthi, R., Sundram, K Tan, Y.A. (2000). Chemistry and biochemistry of palm oil, *Progress in Lipid Research*, Vol. 39, No. 6, pp. 507-558.

Schoenmakers, P.J. Uunk, L.G. (1987). Supercritical fluid chromatography-recent and future developments. *European Chromatography News*, Vol. 1, No. 3, pp. 14-22.

Smith, R.M. (ed.) (1988). Supercritical Fluid Chromatography, Roy Soc Chem, London.

Soares, B.M.C. Gamarra, F.M.C. Paviani, L.C. Gonçalves, L.A.G. Cabral, F.A. (2007). Solubility of triacylglycerols in supercritical carbon dioxide, *J. Supercritical Fluids*, Vol. 43, No. 1, pp. 25-31.

Sow, HP. (1979). Modification in the chemical composition of palm oil by physical, chemical and biochemical treatment, PhD. thesis, Universiti Sains Malaysia, Malaysia.

Tan, T.J. Jinap, S. Kusnadi A.E. Abdul Hamid, N.S. (2008). Extraction of cocoa butter by supercritical carbon dioxide: optimization of operating conditions and effect of particle size, *J. Food Lipids*, Vol. 15, No. 2, pp. 263-276.

Upnmoor, D. Brunner, G. (1989). Retention of acidic and basic compounds in packed column supercritical fluid chromatography, *Chromatographia*, Vol. 28, No. 9-10, pp. 449-454.

Upnmoor, D. Brunner, G. (1992). Packed column supercritical fluid chromatography with light scattering detection. I. Optimization of parameters with a carbon dioxide/methanol mobile phase, *Chromatographia*, Vol. 33, No. 5-6, pp. 255-260.

Weber, W. Petkov, S. Brunner, G. (1999). Vapor liquid equilibria and calculations using the Redlich-Kwong-Aspen-equation of state for tristearin, tripalmitin, and triolein in CO_2 and propane, *Fluid Phase Equilibria*, Vol. 158, pp. 695-706.

Wenclawiak, B. (ed.) (1992). Analysis with Supercritical Fluids: Extraction and Chromatography, Springer, Berlin Heidelberg New York London Paris Tokyo Hong Kong Barcelona Budapest.

White, C.M. (ed.) (1988). Modern Supercritical Fluid Chromatography, Hüthig, Heidelberg Basel New York.

Yamauchi, Y. Saito, M. Hondo, T. Senda, M. Milet, J.L. Castiglioni, E. (1988). Coupled supercritical fluid extraction-supercritical fluid chromatography using pre-concentration/separation column: Application to fractionation of lemon peel oil. In: Perrut, M. (ed.) *Proceedings of 2nd International Symposium on Supercritical Fluids*.

Yamauchi, Y. Saito, M. (1990). Fractionation of lemon-peel oil by semi-preparative supercritical fluid chromatography, *J. Chromatography A*, Vol. 505, No. 1, pp. 237-246.

2

Interaction Parameters of Surfactant Mixtures by Inverse Gas Chromatography

Eleuterio Luis Arancibia[1], Pablo C. Schulz[2] and Susana M. Bardavid[1]
[1]*INQUINOA - CONICET- FACET – UNT- Tucumán,*
[2]*INQUISUR – CONICET –FQ –UNS – Bahia Blanca,*
Argentina

1. Introduction

Inverse Gas Chromatography (IGC) is an accurate and versatile technique for determining a variety of properties of different materials (Voelkel, A. et al., 2009). IGC has also shown to be an effective way to determine thermodynamic properties in polymer materials, especially to obtain Flory-Huggins interaction parameters, which can be used to calculate energy densities interaction (B_{23}) in polymer-polymer systems. These parameters have been employed for evaluating thermodynamic miscibility in polymer mixtures (Deshpande, et al., 1974; Lezcano, et al., 1995).

The major difficulty in those determinations relies on the dependence of the parameters obtained on the solvents used as probe solutes. In spite of the fact that the relations obtained from interaction studies through IGC can be applied to miscible systems, the studies in systems that present partial or total inmiscibility have been made in several polymer systems (Du, et al., 1999; Zhikuan, C. & Walsh, D.J., 1983).

The relation from which B_{23} data are obtained describes a ternary system as a simple balance of the corresponding binary ones. It is well known their small contribution to this relation (Etxeberria, et al., 1994, 1995). Besides, the interaction parameter is several orders of magnitude smaller than that of the retention property determined by IGC. An attempt to reduce the uncertainty in the B_{23} parameter has led to establish an adequate selection of probe solutes (Etxeberria et al., 2000). These authors have also demonstrated that the results obtained from IGC as well as from static technique are similar when CO_2 is used as a probe solute.

As indicated by several authors, the polymer-polymer interaction parameter determined by IGC shows a clear dependence on the solvent used as a probe. In order to solve this problem, different methods have been proposed (El-Hibri, et al., 1988; Etxeberria, et al., 1994, 2000; Farooque, & Deshpande, 1992; Huang, 2003a, 2003b; Prolongo, et al., 1989; Zhao, & Choi, 2001).

Farooque and Deshpande's proposed methodology (Farooque, & Deshpande, 1992) has allowed the determination of reliable polymer-polymer interaction parameters by IGC. Huang's later proposal (Huang, 2003a, 2003b) based on a methodology similar to the latest has been applied to several systems and has been compared with the previous methodology using retention data in polymer materials (Benabdelghani, et al., 2006).

In the case of liquid crystals, they have been thoroughly investigated as stationary phases by gas liquid chromatography (GLC) from an analytical point of view (Janini, et al., 1975; Janini, et al., 1980; Martire, et al., 1968). They have also been used in studies of solution thermodynamic, by inverse gas chromatography (IGC) of solutes in liquid crystals (CL) and in polymer liquid crystals (CLP) at infinite dilution (Chow & Martire, 1971; Romannsky & Guillet, 1994; Romannsky, et al., 1994; Shillcock & Price, 2004).

Huang et al. (Huang, et al., 2007) has studied the solution properties of a series of organic solutes in liquid crystals using IGC through activity coefficients and free energy transference of solute between mesophases and an isotropic phase. In these systems, constant values of free energy of transference have been obtained applying lattice's model of polymer solutions (Flory, 1953) and Flory's treatment of liquid crystals (Flory, 1956; Flory, 1984). The treatment of the data obtained via IGC in order to get thermodynamic properties using Flory's classic model has been more frequently applied to mesophases than to Flory's liquid crystal model. In the derivation of this model, the expression that corresponds to the non-combinatory contribution has been discarded (Abe & Flory, 1978; Flory & Abe, 1978); thus, the interaction parameter $\chi_{1,2}$ between probe solute and stationary phase, that represents the energy exchange of interactions, has not been included. This might be one of the reasons because the interaction parameter is obtained preferably by Flory's classic model of polymers.

Since the probe solute is at near infinite dilution, the order of the liquid crystal phase is not destroyed by the probe solute inclusion. The accessibility of the probe molecules in a mesophase could be limited by its ordering. This means that the main interaction occurs with the hydrocarbon chains of the bilayers and the interactions are mainly of dispersive character. We can estimate the degree of interaction solute – stationary phase by Flory Huggins' interaction parameter obtained by IGC using surfactants as stationary phase.

The amphiphilic molecular structure of surfactants has a significant influence in the crystals phase structure. The packing of the surfactants is produced so that the liphophilic groups of the different molecules are associated with every lipophilic region. Their hydrophilic groups are equally associated within the polar region. In this way, they form the so called bilayers which are usually formed in crystals of simple or double hydrocarbon chain surfactants. Some surfactant crystals do not melt in a liquid phase directly but go through anhydrous liquid crystal phases (thermotropic liquid crystals) before reaching the isotropic liquid state. The liquid crystals of anhydrous surfactants are thermotropic since they result only from the temperature increase on anhydrous crystals (Laughlin, 1994).

This chapter is part of a series of works done in this laboratory, where several surfactant properties such as solubility parameters, surfactant-surfactant interaction parameters in several systems made up by cationic surfactants of different hydrocarbon length chain have been determined by IGC (Bardavid, et al., 2003, 2007; Proverbio, 2003; Schaefer, 2008).

The vast majority of the works related to surfactant mixtures have been made at low concentration in aqueous solutions. The mixing of different types of surfactants gives rise to synergies that provide the opportunity to optimize product performance. For surfactant mixtures the characteristic phenomena are the formation of mixed monolayers at the interface and mixed micelles in the bulk solution. In such solutions, adsorption behaviour, aggregates' microstructure, and rheological properties can be manipulated to tailor the properties of the different products.

In ionic anhydrous surfactant systems, the structures of liquid crystals and crystals are based on the simultaneous fulfillment of two kinds of interactions: van der Waals interactions in the hydrocarbon bilayers and the electrostatic interactions in the ionic bilayers. Sometimes, steric interactions can appear in the hydrocarbon bilayers, and polar interactions or hydrogen bonds can appear in the ionic bilayers (Schulz et al., 1996). A combination of these interactions can be especially disclosed in mixed amphiphiles systems, and their study can lead to a better understanding of their influence in the formation and stability of the microstructures mentioned above.

The cationic surfactants mixed systems are becoming more important and in the future additional complex formulation a and multiple technological products will be required. The extension to a thermodynamic approximation of multicomponents including additional phenomena like solubility will allow the establishment of more complex systems (Holland & Rubingh, 1990). There is not much information about studies of cationic surfactant mixture phases (Varade, et al., 2008), or the miscibility of pure surfactant mixtures (Bardavid, et al., 2007, 2010) that allow resemblance to the behaviour of surfactant mixtures interactions in mixed micelles.

In this chapter we present the results obtained from the study of miscibility of cationic surfactant mixtures of three systems made up by mixtures of surfactants of equal polar head and different hydrocarbon chains by determining surfactant-surfactant interaction parameters through IGC. The implementation of this technique has also allowed us to analyze the use of two methodologies of measurement in order to obtain the parameter B_{23} for cationic surfactants and to contrast them with the ones obtained in polymeric materials. With this work we hope to enlarge the information about the behaviour of pure surfactant mixtures, and analyze the non-ideality degree in the mixtures and its possible causes.

2. Experimental

2.1 Materials

Dodecylpyridinium chloride (DPC), Hexadecylpyridinium bromide (Cetylpyridinium bromide (CPB)), Dodecyltrimethylammonium bromide (DTAB), Octadecyl trimethylammonium bromide (OTAB), (Aldrich, analytical grade, USA) Didodecyldimethylammonium bromide (DDAB) and Dioctadecyldimethylammonium bromide (DODAB) (Sigma, analytical grade, USA) were used as received. All probe solutes were chromatographic quality or reagent grade and were used without further purification.

2.2 Differential Scanning Calorimetry (DSC)

DSC was performed on a Perkin Elmer DSC 6 calorimeter, between 293 and 523 K, with a scanning rate of 10 degree min^{-1} and using samples of 5–10 mg for pure surfactants and 10–15 mg for materials collected over chromatography support. The instrument was calibrated with indium.

2.3 Inverse Gas Chromatography (IGC)

Pure surfactants and their mixtures were used as stationary phase and deposited on Chromosorb W or G, AW, 60/80, which was employed as solid support. The column filler was prepared using methanol as a solvent in a rotary evaporator under a flow of dry nitrogen and was kept in a dry atmosphere before filling the columns (stainless steel pipes). The column was loaded and conditioned for 1 h at 363 K under a flow of carrier gas. The amount of stationary phase on the support was determined by calcinations of about one gram of material. The data employed in the specific retention volume computation were obtained by using a column 100 cm long, 1/4 inch external diameter, and the packing characteristics are included in Table 1.

System DDAB (2) – DODAB (3)			
Stationary phase	Mass packing (g)	Loading(w/w) (%)	Weight fraction (w_2)
DDAB (2) + DODAB (3)	7.2653	9.12	0.0000
DDAB (2) + DODAB (3)	7.6432	9.22	0.1643
DDAB (2) + DODAB (3)	7.8634	9.23	0.3281
DDAB (2) + DODAB (3)	7.0268	9.53	0.5037
DDAB (2) + DODAB (3)	7.0906	9.60	0.7325
DDAB (2) + DODAB (3)	7.8856	9.11	1.0000
System DPC (2) – CPB (3)			
Stationary phase	Mass packing (g)	Loading(w/w) (%)	Weight fraction (w_2)
DPC (2) + CPB (3)	13.2318	7.44	0.0000
DPC (2) + CPB (3)	12.9175	7.82	0.1598
DPC (2) + CPB (3)	12.9056	7.72	0.3461
DPC (2) + CPB (3)	12.6945	7.42	0.5409
DPC (2) + CPB (3)	12.9240	7.39	0.7590
DPC (2) + CPB (3)	12.5835	7.41	1.0000
System DTAB (2) – OTAB (3)			
Stationary phase	Mass packing (g)	Loading(w/w) (%)	Weight fraction (w_2)
DTAB (2) + OTAB (3)	7.2301	10.09	0.0000
DTAB (2) + OTAB (3)	6.7947	10.04	0.1102
DTAB (2) + OTAB (3)	7.4240	10.02	0.3652
DTAB (2) + OTAB (3)	7.3281	9.99	0.5963
DTAB (2) + OTAB (3)	7.2392	11.64	0.8331
DTAB (2) + OTAB (3)	7.0746	10.03	1.0000

Table 1. Column loading data and the weight fraction at the different mixtures.

The retention time measurement for each solute was performed with a Hewlet Packard, HP 6890 series, GC System, equipped with a flame ionization detector (FID). Column temperature was measured in a range between 343.1 and 403.1 K with an Iron-Constantan thermocouple placed in the direct environment of the column. The temperature stability during experiments was ± 0.2 K. The employed solutes were n-hexane, n-heptane, n-octane, n-nonane, cyclohexane, methylcyclohexane, benzene, toluene, ethyl acetate, dichloromethane, trichloromethane and carbon tetrachloride. Nitrogen was used as carrier gas.

Flow rates were measured at the beginning of each experiment with an air-jacketed soap film flowmeter placed at the outlet of the detector. Inlet pressures were measured with a micrometry syringe (trough the injector septum) which was connected to an open branch mercury manometer. To ensure that the results were independent of sample size and flow rate and those measurements were being made at infinite dilution the usual checks were made (Conder & Young, 1978). Solutes were injected with 10 µl Hamilton syringes, as steam in equilibrium with pure liquid. For all the solutes and for all the range of stationary phase concentrations the peaks were symmetric. The injector was kept at 423 K and the detector at 453 K.

Retention times (t_R) were measured with a Chem Station system and the retention specific volumes (V^0_g) were calculated with the following relationship (Conder & Young, 1978):

$$V_g^0 = j\left(\frac{F_f}{w}\right)\left(\frac{273.15}{T_f}\right)(t_R - t_0)\left(\frac{(p_0 - p_w)}{p_0}\right) \tag{1}$$

where j is the James-Martin compressibility correction factor, p_0 represents the outlet column pressure, F_f is the flow rate measured at pressure p_0 and temperature T_f, w is the mass of the stationary phase into the column and p_w is the water vapour pressure at T_f; t_0 is the dead time, which was measured by using the methane peak obtained with the FID.

3. Data reduction

Specific retention volumes were fitted to the equation (Conder & Young, 1978):

$$\ln V_g^0 = -\Delta H_s^0 / RT + constant \tag{2}$$

where ΔH°_s is the sorption heat. The values obtained for ΔH°_s, as well as their respective standard deviations, were calculated using Marquartd-Levenberg's algorithm (Marquartd, 1963) and can be seen in Table 2. Values of standard deviations in ΔH_s smaller than 1.5 % are obtained in the regression of specific retention values vs. 1/T, although most of the values were near 1.0 %.

The meaning of ΔH°_s depends on the physical state of the stationary phase. For a solid, ΔH°_s correspond to the molar adsorption enthalpy. For the liquid mesophase, it was assumed that the solute is dissolved in the stationary phase so ΔH°_s corresponds to the molar solution enthalpy.

The average per cent error values ΔH°_s for DDDAB is 0.88 %, and 0.74% for DODAB. For the DPB-CPB systems the average error % is 0.79 % for DPC and for CPB is 1.20 %. Finally, for the mixed systems DTAB-OTAB the average error in ΔH°_s values is generally bigger than the previous ones and from the calculated values we get a value of 1.38 % for the average error in DTAB and of 1.10 % for the values in OTAB.

System DDDAB-DODAB					
	DDDAB		DODAB		
	ΔH°_{s}	$\pm\sigma$	ΔH°_{s}	$\pm\sigma$	Difference
n-Hexane	28.7	0.3	30.7	0.3	2.0
n-Heptane	33.4	0.4	34.8	0.4	1.4
n-Octane	38.6	0.2	39.3	0.3	0.7
Benzene	33.5	0.2	33.7	0.4	0.2
Toluene	37.5	0.4	37.6	0.2	0.1
Cyclohexane	29.8	0.4	31.1	0.2	1.3
Methylcyclohexane	33.6	0.2	33.6	0.2	0.0
Dichloromethane	33.2	0.3	32.3	0.3	-0.9
Trichloromethane	43.8	0.4	43.4	0.2	-0.4
Carbon tetrachloride	35.5	0.2	34.3	0.2	-1.2
Ethyl acetate	33.3	0.3	32.6	0.1	-0.7
System DPC-CPB					
	DPC		CPB		
	ΔH°_{s}	$\pm\sigma$	ΔH°_{s}	$\pm\sigma$	Difference
n-Hexane	29.5	0.1	27.4	0.5	-2.1
n-Heptane	33.6	0.2	31.8	0.5	-1.8
n-Octane	38.1	0.2	35.9	0.4	-2.2
Benzene	32.9	0.3	30.2	0.3	-2.7
Toluene	37.3	0.2	35.1	0.6	-2.6
Cyclohexane	29.8	0.2	27.2	0.4	-2.6
Methylcyclohexane	32.2	0.3	30.5	0.5	-1.7
Dichloromethane	31.4	0.4	27.1	0.2	-4.3
Trichloromethane	40.4	0.3	35.9	0.3	-4.5
Carbon tetrachloride	33.9	0.3	31.4	0.2	-2.5
Ethyl acetate	31.6	0.4	28.6	0.3	-3.0
System DTAB-OTAB					
	DTAB		OTAB		
	ΔH°_{s}	$\pm\sigma$	ΔH°_{s}	$\pm\sigma$	Difference
n-heptano	35.0	0.5	33.1	0.3	-1.9
n-Octano	39.4	0.2	38.4	0.4	-1.0
n-nonano	43.3	0.7	40.7	0.5	-2.6
Benceno	38.5	0.2	35.2	0.4	-3.3
Tolueno	42.6	0.8	39.0	0.5	-3.6
Cyclohexane	30.9	0.6	28.7	0.3	-2.2
Methylcyclohexane	33.5	0.6	31.3	0.4	-2.2
Dichloromethane	33.8	0.2	30.8	0.3	-3.0
Trichloromethane	47.2	0.8	43.1	0.4	-4.1
Carbon tetrachloride	34.7	0.8	32.5	0.4	-2.2
Ethyl acetate	33.6	0.3	29.7	0.3	-3.9

Table 2. Solution heat ($kJ.mol^{-1}$) and standard deviations for of the surfactant mixtures.

Activity coefficients at infinite dilution in terms of mole fraction were obtained by the following expression (Price et al., 2002):

$$\ln \gamma_i^{\infty} = \ln \frac{273.15R}{V_g^0 p_1^0 M_2} - \frac{p_1^0 (B_{11} - V_1)}{RT} \tag{3}$$

where M_2 stands for molar mass of surfactant, p_1^0, V_1 stands for vapour pressure and molar volume of pure solute. B_{11} is the second virial coefficient for solute–solute interactions.

Patterson (Patterson et al., 1971) suggested using the weight fraction, w, in which case equation (3) may be replaced by expression:

$$\ln \Omega_i^{\infty} = \ln \frac{273.15R}{V_g^0 p_1^0 M_1} - \frac{p_1^0 (B_{11} - V_1)}{RT} \tag{4}$$

where M_1 is the molar mass of the solute.

The Flory-Huggins theory of non-athermal solutions gives:

$$\ln a_1 = (\ln a_1)_{comb.} + (\ln a_1)_{noncomb.} = \left(\ln \phi_1 + \left(1 - \frac{1}{r}\right)\phi_i \right) + \chi \phi_i^2 \tag{5}$$

where the volume fraction, ϕ, is defined in terms of the specific volumes.

For the limiting case in which ϕ_1 tends to unity, using the equations (4) and (5), the probe solute-surfactant Flory-Huggins interaction parameter, χ_{1i}, can be calculated from the specific retention volumes, V^o_g, by the expression (Conder & Young, 1978; Deshpande, et al., 1974):

$$\chi_{1i}^{\infty} = \ln \left(\frac{273.15Rv_i}{V_g^0 p_1^0 V_1} \right) - \left(\frac{p_1^0 (B_{11} - V_1)}{RT} \right) - \left(1 - \frac{V_1}{V_i}\right) \tag{6}$$

where v_i stands for the specific volume of the surfactant experimentally measured in the laboratory.

The vapour pressures were computed using Antoine equations and the coefficients were taken from Riddick, Bunger and Sakano (Riddick, et al., 1986). The solute densities at different temperatures were estimated from Dreisbach's compilation (Dreisbach, 1955). The second virial coefficient of the solutes was calculated by Tsonopoulos's correlation using critical constants tabulated in Reid et al. (Reid et al., 1986).

When the stationary phase is a surfactant mixture, Equation (7) allows to determine the ternary probe solute (1)-surfactant (2)-surfactant (3) interaction parameter, $\chi_{1(23)}$, assuming an additive specific volume for the surfactant mixture, $v_m = w_2 v_2 + w_3 v_3$ where w_i is the weight fraction of surfactant i in the mixture (Deshpande, et al., 1974).

$$\chi_{1(23)}^{\infty} = \ln\left(\frac{273.15R(w_2v_2 + w_3v_3)}{V_g^0 p_1^0 V_1}\right) - \left(\frac{p_1^0(B_{11} - V_1)}{RT}\right) - \left(\phi_2\left(1 - \frac{V_1}{V_2}\right) + \phi_3\left(1 - \frac{V_1}{V_3}\right)\right) \quad (7)$$

where ϕ_i stands for the volume fraction for i component in the stationary phase. V_i is the molar volume of component i in the mixture.

On the contrary, assuming the Scott-Tompa approximation (Tompa, 1956), which describes a ternary system as a simple balance of the corresponding binary systems it is possible to calculate the surfactant-surfactant interaction parameter, χ_{23}, by:

$$\chi_{1(23)}^{\infty} = \phi_2\chi_{12}^{\infty} + \phi_3\chi_{13}^{\infty} - \phi_2\phi_3\chi_{23}\left(\frac{V_1}{V_2}\right) \quad (8)$$

As it has been indicated by different authors, the polymer-polymer interaction parameter determined by IGC shows a clear dependence on the solvent used as a probe. In order to solve this problem, different methods have been proposed. The Farooque and Deshpande (Farooque, & Deshpande, 1992) and the Huang (Huang, 2003a, 2003b) methodologies will be applied to retention data obtained from the use of surfactant anhidrous mixtures in order to carry out a comparative analysis of the behaviour of these methodologies in the determination of surfactant-surfactant interaction parameters.

Farooque and Deshpande (Farooque, & Deshpande, 1992) methodology gives a reliable true interaction parameter after a rearrangement of Eq. (8):

$$\frac{\left[\chi_{1(23)}^{\infty} - \chi_{13}^{\infty}\right]}{V_1} = \left[\frac{\phi_2\left(\chi_{12}^{\infty} - \chi_{13}^{\infty}\right)}{V_1}\right] - \phi_2\phi_3\left[\frac{\chi_{23}}{V_2}\right] \quad (9)$$

A plot of the left side of this expression versus the first term of the right-hand side yields a lineal function from whose slope ϕ_2 can be calculated and from the intercept χ_{23} can be obtained. The physical meaning of the slope was interpreted in terms of an effective average column composition that the solutes are probing.

Huang (Huang, 2003a, 2003b)et. al. have proposed an alternative rearrangement of Eq.(8):

$$\frac{\chi_{1(23)}^{\infty}}{V_1} = \left[\frac{\phi_2\chi_{12}^{\infty} + \phi_3\chi_{13}^{\infty}}{V_1}\right] - \phi_2\phi_3\left[\frac{\chi_{23}}{V_2}\right] \quad (10)$$

A linear plot can be obtained from the left-hand side vs. first term of right-side of Eq.(10) allowing that the interaction parameter can be obtained.

In both methodologies, if the conditions given by Al-Saigh and Munk (Al-Saigh & Munk, 1984) are obeyed, the surfactant-surfactant interaction parameter can be calculated through the specific retention volume without calculating the individual parameter.

The values of the surfactant-surfactant interaction parameters can be analyzed as χ_{23}/V_2 or as χ_{23} when multiplied by V_2, or as the equivalent quantity $B_{23} = RT\ (\chi_{23}/V_2)$ (in $J.cm^{-3}$) called energy density.

4. Results and discussion

4.1 Surfactants as stationary phases

In this chapter both DSC and IGC were used to confirm the stationary phase stability with temperature. The phase transition temperatures were determined with DSC on pure surfactants between 293 and 523 K. The same technique was also employed to analyze the thermal behaviour of both surfactants deposited over the solid support in a 10 % (w/w) percentage, approximately.

Chow and Martire (Chow & Martire, 1971) compared IGC and DSC studies on two azoxy liquid crystals and reported no measurable adsorption effects from the interface above a film thickness of 100 nm. Witkiewicz (Rayss, J. et al., 1980) reported surface orientation effects up to depth of 2 nm, but in a later work reported constant specific retention volumes above a stationary phase loading of above 5% (Marciniak & Witkiewicz, 1981). Zhou et al. (Zhou, 1994) in the GC and IR study of liquid crystal deposited on different types of silica have shown that for a percentage under of 7 % of the stationary phase loading, the ln Vg vs 1/T plot did not show discontinuity. The loading used in this work was near of 10 % on Chromosorb W in all the cases.

The retention diagram of ln Vg vs 1/T for solute probes in DODAB and DDAB coated on Chromosorb W, NAW, 60/80, is shown in Figure 1 (Bardavid, et al., 2007). For both surfactants, specific retention volumes were obtained between 303 K and 423 K. On heating the crystalline solid, the retention decreases until the crystalline solid-to-liquid crystal transition is reached. Then there is a large increase in retention, which once the system phase change finished, decreases with increasing temperature. For DDAB and DODAB changes in retention are observed at 343.1 K and 358.1 K respectively.

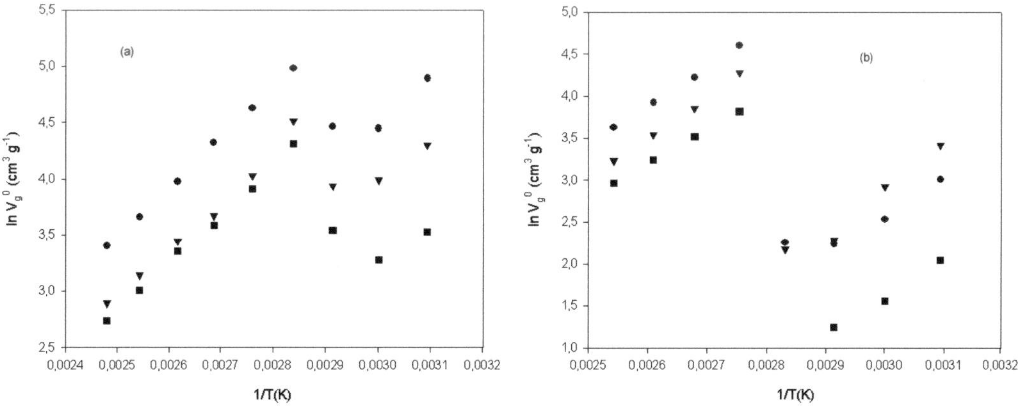

Fig. 1. The retention diagram for solute probes in DDAB (a) and DODAB (b) coated on Chromosorb W. Solutes: (●), toluene; (■), carbon tetrachloride; (▼), n-octane.

DSC measurements of the phase transition temperature for DDAB y DODAB have been described in the literature (Schulz, et al., 1994, 1998). The experimental measurements for DDAB (Bardavid, et al., 2007), DSC analysis shows a temperature transition between the solid phase and liquid crystal mesophase through two peaks at 336.8 K and 349.2 K. These values were coincident with those obtained in bibliography (Schulz, et al., 1994). The first peak corresponds to the melting of hydrocarbon tails of DDAB and that at 349.2 K corresponds to the melting of the DDAB polar heads bilayer giving an anhydrous lamellar liquid crystal which in some circumstances (i.e., when lamellae are parallel to the slide surface) appears as pseudo-isotropic. There is another phase transition at 445.9 K that could be the transition to isotropic liquid. DSC values in the literature have been informed up to 445.9 K.

DSC analysis for pure DODAB shows a phase transition at 361.8 K (Bardavid, et al., 2007). According to the literature (Schulz, et al., 1994) this transition corresponds to the melting of DODAB crystals to a liquid which was named a pseudoisotropic liquid (Schulz, et al., 1998). We have not detected in our thermogram the hydrocarbon tails transition temperature, perhaps because both transitions (i.e., the melting of the polar and the apolar layer) occur at almost the same temperature giving an overlapping of their peaks; but we have detected another transition temperature at 439.3 K that we considered as the transition from mesophase to isotropic liquid. The transitions for DDAB and DODAB in literature (Schulz, et al., 1994, 1998) were studied up to smaller temperatures than ours. DDAB and DODAB supported on Chromosorb W NAW show a slight displacement toward smaller temperatures with respect to pure surfactants. There is previous information about discrepancies in the results when liquid and supported samples are analyzed by DSC, with lower values of transition temperature for the last ones (Shillcock & Price, 2003).

The systems DPC and CPB, either pure or mixed, are lamellar mesophases at the working temperature (Laughlin, 1990). The obtained results trough DSC in pure surfactant samples for the change from crystal phase to liquid crystal phase is one peak at 343.1 K and others two peaks at 436.1 K and 513.1 K for DPC (Bardavid, et al., 2011).

In the case of CPB one peak is at 339.4 K and the other two at 421.7 K and 518.1 K (Bardavid, et al., 2011). When the surfactants are deposited on chromatographic support (Chromosorb G, AW, 60/80) the values obtained through DSC for the phase changes are at

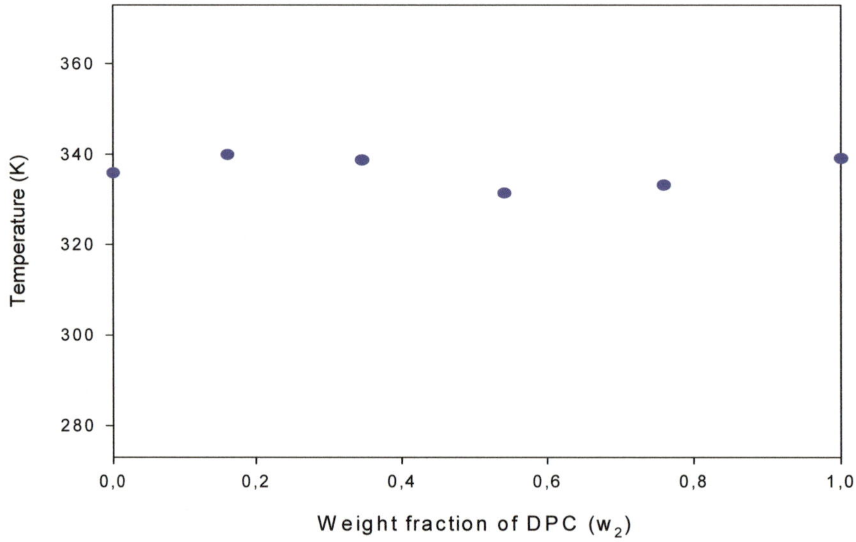

Fig. 2. Phase transition temperature for the surfactant mixtures DPC-CPB deposited on the solid support as function of the weight fraction of component 2 (DPC).

339.1 K for DPC and at 335.9 K for CPB. In Figure 2 we have included the values of phase transition temperature for the surfactant mixtures deposited on the solid support as function of the weight fraction of component 2 (DPC) (Bardavid, et al., 2011).

When the studied system consists of DTAB and OTAB either pure or mixed, deposited on Chromosorb W, produced lamellar mesophases at work temperature (Laughlin, 1990). DSC experiments showed the phase changes from crystal to liquid crystal occurring at 372.5 K for DTAB and at 378.1 K for OTAB (Bardavid, et al., 2010). IGC measurements of the retentive behavior of n-octane and toluene between 338.1 and 423.1 K indicate (in the ln Vg vs. 1/T plot) retention changes at 368.1 K for DTAB and 373.1 K for OTAB. The anhydrous crystal to lamellar phase transition in pure dodecyltrimethylammonium chloride (DTAC) occurs at about 356.1 K (Blackmore & Tiddy, 1990). Taking into account the effect of changing the counterion, the agreement is good.

The DSC technique is very accurate in determining the pure component properties and the IGC results can be used as supplement to the DSC results in the case of surfactant deposited on solid support. The values of transition temperatures obtained by IGC are always lower than those obtained by DSC, and even more when the very start detection method is used to obtain the transition temperature by means of gas chromatography (Benabdelghani, et al., 2006; Nastasovic & Onjia, 2008; Shillcock & Price, 2003).

According to these results, it is possible to point out that both pure surfactants and their mixtures appear as a lamellar mesophase at the working temperature (388-403 K). Thus, we can infer that the probe solutes dissolve in a stationary phase formed by an ordered structure of lamellar mesophase.

4.2 Interaction parameter in mixtures

The systems studied, which are made up by cationic surfactants deposited on Chromosorb W or G, being either pure or mixed, appear as lamellar mesophase at work temperature (Laughlin, 1990) and their temperatures of transition phases have been determined by DSC and IGC. In two systems we have used Chromosorb W, 60/80, with a charge close to 10 % of stationary phase, and when Chromosorb G, 60/80 was used, the percentage of charge of stationary phase was close to 7 %. These values are considered to be adequate for this type of chromatographic support (Conder & Young, 1978; Nastasovic & Onjia, 2008). To ensure that the results were independent of sample size and flow rate and since measurements were being made at infinite dilution, the usual checks were made (Conder & Young, 1978).

As shown by Etxabarren et al. (Etxabarren, et al., 2002), in polymeric materials the polymer-probe solute interaction parameters depend on the polymer molecular mass in intermediate concentration zones, and this dependence vanishes at very high concentration as in the case presented in IGC. These conclusions have lead us to analyze the behaviour in surfactant mixtures which present molecular mass that are much lower than those of the polymers with Tompa's approximation (Tompa, 1956) to ternary systems in Flory Huggins's theory.

As pointed out in the introduction, the measurements made in liquid state through IGC allow the determination of thermodynamic properties (Deshpande, et al., 1974), especially of interaction parameters that play an important role in determining the miscibility of mixtures. The behaviour of surfactant mixtures has been determined by IGC allowing the experimental determination of surfactant-surfactant interaction parameters in systems with different characteristics (Bardavid, et al., 2007, 2010, 2011).

Negative values of B_{23} are indicative of attractive interactions and hence higher miscibility. On the contrary, positive values of B_{23} would indicate repulsive interactions between the two polymers and they are related to immiscibility (Al-Saigh & Munk, 1984; Benabdelghani, et al., 2006; Deshpande, et al., 1974; DiPaola-Baranyi &. Degre, 1981; Etxeberria et al., 2000; Shi & Screiber, 1991). From a theoretical point of view, this parameter should be constant with the concentration. Nevertheless, it has frequently been found that its value changes with concentration (Etxeberria, et al., 1994).

Through IGC (Bardavid, et al., 2007) we have studied the system DDAB (didodecyldimethyl ammonium bromide) and DODAB (dioctadecyldimethyl ammonium bromide), which are two twin tailed surfactants, the system DTAB (Dodecyltrimethyl ammonium bromide) and OTAB (Octadecyltrimethylammonium bromide) (Bardavid, et al., 2010) and the system DPC (Dodecylpyridinium chloride) and CPB (Hexadecylpyridinium bromide) (Bardavid, et al., 2011). The values of B_{23} obtained in the former were positive in all the range of concentration and at all temperatures. They were also indicative of high inmiscibility and they coincided with the behaviour of these surfactant mixtures in aqueous solutions (Feitosa, et al., 2006). This is not surprising because the structure of anhydrous lamellar liquid crystals formed by melting the crystals must fulfill the same conditions that lamellar mesophases formed in aqueous solutions, i.e., polar headgroups must be in polar layers and the chains in apolar bilayers.

In the system DTAB and OTAB most of the values in B_{23} are negative for each studied temperature. Negative values of B_{23} suggest that the interaction between surfactants is more favourable as DTAB concentration increases. There are some precedents in aqueous solutions in which these surfactants form mixed micelles (Akisada, et al., 2007; Schulz, et al., 2006).

In the system DPC and CPB all the B_{23} values are positive in the range of concentration and temperature studied. These values are lower than the B_{23} values obtained in the DDAB and DODAB system. To our knowledge, studies of mixtures DPC and CPB, neither pure nor aqueous solution, have not been carried out.

In this mixture of surfactants deposited on a solid support, an ordering can be expected due to the fact that the polar head would be directed to a solid surface. There is some indication that the pyridinium head group may interact specifically with surface sites, most likely by hydrogen bonding (Fuerstenau & Jia, 2004). Thus it is expected that the repulsion between the counter ions and the π-electron cloud on the pyridinium ring together with lower charge on nitrogen atom is responsible for high α-value of such surfactants (α being the micelle ionization degree), higher than their trialkylammonium counter-parts (Bhat, et al., 2007). This would explain the difference in the behaviour in these systems.

As usual, in polymer mixtures increasing the temperature the interactions become weaker; thus the values of B_{23} become more positive at higher temperatures (Benabdelghani, et al., 2006). A similar behaviour can be observed in the systems with the surfactants studied.

4.3 Farooque-Deshpande and Huang methodology

Farooque and Deshpande (Farooque, & Deshpande, 1992) and Huang's (Huang, 2003a, 2003b) methodologies have been shown to be effective and reliable to determine the interaction parameter χ_{23} in polymers by IGC, since these parameters were questioned because of their dependence on the probe solutes used.

We have applied the thermodynamic relations determined by IGC for polymer mixtures to anhydrous surfactant mixtures and have proved that the results obtained show a behaviour that is coherent among the systems studied. Similarly, we have proved that the results are coherent with some results obtained in aqueous solutions of cationic surfactants. It's worth noting that the equations used in the measurement of B_{23} would be only valid for miscible mixtures, although they have been applied successfully to several systems that show immiscibility (Du, et al., 1999; Zhikuan, & Walsh, 1983). Benabdelghani et al. (Benabdelghani, et al., 2006) have carried out an analysis of both methodologies in a study of phase behaviour of poly(styrene-co-methacrylic acid)/poly(2,6-dimethyl-1,4-phenylene oxide) through IGC of polymer mixtures.

They have concluded that both methods show similar interaction parameter values and that both can be considered as reliable to determine the true polymer mixture parameters.

The possibility to count with experimental data in three binary systems of anhydrous surfactants, their molar volume and to devise a method to classify miscibility in mixtures has led us to apply both methodologies to surfactant mixtures. Besides, this allows us to prove if both methods can be used in these types of substances.

In Table 3 we have included the values obtained from the intercept, its errors and the correlation coefficients with the lineal fitting of the data applied to the equations (9 and 10) using both methodologies for the system DDAB - DODAB. This system presents positive values of B_{23} that indicate the presence of immiscibility in this mixture.

Several observations can be made on the obtained values. First, the interception values in this system (DDAB - DODAB), are nearly equal for both methodologies. Using Huang's methodology the errors are the double or larger. The correlation coefficients of the lineal regression are excellent in Huang's methodologies in contrast to Farooque y Deshpande's (F-D) (Farooque, & Deshpande, 1992) which are good.

These results are similar to those obtained from the comparison between both methods made by Benabdeghani et al. (Benabdelghani, et al., 2006) in the system Poly(styrene-co-methacrylic acid) - Poly(2,6-dimethyl-1,4-phenylene oxide), which have led them to conclude that both methods are reliable to calculate interaction parameter values in polymer mixtures with miscible and immiscible regions.

In Table 4 we have included the system DPC - CPB, which also presents positive values of B_{23} that indicate the presence of immiscibility in this mixture, where the situation is different. In general, the intercept values are lower using Huang's method and errors are higher (five times or more) than those determined by F-D method. Something peculiar appears at a minor percentage of component 3 (CPB). At all temperatures the obtained intercept values are similar and the errors obtained by Huang are slightly higher.

363.1 K	F - D			Huang		
w_2(DDDAB)	ord.10^4	$\pm\sigma.10^5$	r^2	ord.10^4	$\pm\sigma.10^5$	r^2
0.1643	18.64	1.28	0.9779	18.91	3.40	0.9999
0.3281	19.98	2.05	0.9902	19.61	2.51	0.9999
0.5037	18.34	4.90	0.9796	17.30	6.15	0.9996
0.7325	17.06	4.51	0.9926	15.04	3.63	0.9998
Average		3.18			3.92	
373.1 K	F - D			Huang		
w_2	ord.10^4	$\pm\sigma.10^5$	r^2	ord.10^4	$\pm\sigma.10^5$	r^2
0.1643	21.11	2.95	0.9260	21.34	5.16	0.9997
0.3281	23.54	3.11	0.9812	23.41	5.10	0.9997
0.5037	21.81	3.83	0.9900	20.57	6.31	0.9996
0.7325	20.62	4.63	0.9931	19.40	6.05	0.9996
Average		3.63			5.65	
383.1 K	F - D			Huang		
w_2	ord.10^4	$\pm\sigma.10^5$	r^2	ord.10^4	$\pm\sigma.10^5$	r^2
0.1643	25.95	1.80	0.9538	27.17	7.16	0.9996
0.3281	27.86	2.12	0.9908	28.30	5.14	0.9998
0.5037	26.47	3.80	0.9912	25.27	7.68	0.9995
0.7325	24.48	7.81	0.9801	25.68	13.69	0.9981
Average		3.88			8.42	
393.1 K	F - D			Huang		
w_2	ord.10^4	$\pm\sigma.10^5$	r^2	ord.10^4	$\pm\sigma.10^5$	r^2
0.1643	32.59	4.31	0.8580	35.00	8.24	0.9995
0.3281	33.76	3.41	0.9835	35.32	5.09	0.9998
0.5037	33.36	4.21	0.9906	34.60	7.04	0.9996
0.7325	30.96	6.12	0.9910	33.34	10.61	0.9990
Average		4.51			7.74	
403.1 K	F - D			Huang		
w_2	ord.10^4	$\pm\sigma.10^5$	r^2	ord.10^4	$\pm\sigma.10^5$	r^2
0.1643	40.50	3.48	0.8936	42.28	12.64	0.9991
0.3281	41.19	5.09	0.9684	42.45	12.18	0.9991
0.5037	39.76	6.09	0.9847	40.44	13.24	0.9988
0.7325	37.40	7.75	0.9886	39.43	18.72	0.9974
Average		5.60			14.19	

Table 3. Intercepts (ord), standard deviations (σ), and correlation coefficients (r^2) in the system DDDAB - DODAB.

348.1 K	F - D			Huang		
w_2(DPC)	ord.10^4	$\pm\sigma.10^5$	r^2	ord.10^4	$\pm\sigma.10^5$	r^2
0.1598	9.41	3.46	0.9846	4.94	11.45	0.9990
0.3461	8.48	3.84	0.9944	3.17	11.46	0.9990
0.5409	8.66	4.85	0.9958	3.54	10.12	0.9990
0.7590	12.07	3.81	0.9984	10.14	4.40	0.9999
Average		4.08			10.76	
358.1 K	F - D			Huang		
w_2	ord.10^4	$\pm\sigma.10^5$	r^2	ord.10^4	$\pm\sigma.10^5$	r^2
0.1598	15.30	5.39	0.9557	12.34	15.39	0.9982
0.3461	14.80	3.60	0.9950	8.76	12.38	0.9989
0.5409	13.36	3.82	0.9975	7.14	11.42	0.9991
0.7590	17.74	3.50	0.9986	16.50	3.87	0.9999
Average		4.08			10.76	
368.1 K	F - D			Huang		
w_2	ord.10^4	$\pm\sigma.10^5$	r^2	ord.10^4	$\pm\sigma.10^5$	r^2
0.1598	18.28	3.91	0.9846	10.56	23.39	0.9964
0.3461	18.51	2.99	0.9952	10.23	17.84	0.9979
0.5409	16.67	3.84	0.9976	7.39	16.37	0.9982
0.7590	21.93	2.86	0.9991	19.57	3.07	0.9999
Average		3.40			15.17	
	F - D			Huang		
378.1 K	ord.10^4	$\pm\sigma.10^5$	r^2	ord.10^4	$\pm\sigma.10^5$	r^2
0.1598	20.50	7.51	0.9676	7.33	47.29	0.9874
0.3461	21.42	4.81	0.9942	7.53	35.96	0.9925
0.5409	18.90	4.78	0.9974	2.65	32.48	0.9939
0.7590	23.85	1.81	0.9997	17.53	8.27	0.9995
Average		4.73			31.00	

Table 4. Intercepts (ord), standard deviations (σ), and correlation coefficients (r^2) in the system DPC- CPB.

In Huang's method the correlation coefficients for all percentages and at all temperatures are greater than 0.999 and in the method F – D the value of these coefficients is between 0.95 and 0.999.

In Table 5, the values obtained from the intercept, their error and correlation coefficients from the lineal fitting of the data for the system DTAB - OTAB are included. This system presents a partial miscibility in accordance with the values of B_{23} obtained by F – D, the intercept values determined by both methodologies present major differences between the systems studied; all the values determined by F - D are negative except for the first values at the last two temperatures.

388.1 K	F - D			Huang		
w_2(DTAB)	ord.10^4	±σ.10^5	r^2	ord.10^4	±σ.10^5	r^2
0.1102	-6.31	43.51	0.9763	12.45	81.23	0.9548
0.3652	-13.52	43.27	0.9940	0.45	80.88	0.9564
0.5963	-15.20	41.07	0.9745	-2.81	74.23	0.9632
0.8331	-13.89	45.15	0.9523	-1.72	80.22	0.9566
Average		43.25			79.14	
393.1 K	F - D			Huang		
w_2	ord.10^4	±σ.10^5	r^2	ord.10^4	±σ.10^5	r^2
0.1102	-1.20	39.70	0.9731	18.84	80.74	0.9548
0.3652	-9.62	39.74	0.9948	5.79	79.42	0.9575
0.5963	-11.31	41.13	0.9732	2.66	82.05	0.9549
0.8331	-9.98	41.82	0.9632	3.31	81.04	0.9558
Average		40.60			80.81	
398.1 K	F - D			Huang		
w_2	ord.10^4	±σ.10^5	r^2	ord.10^4	±σ.10^5	r^2
0.1102	2.44	36.96	0.9716	22.97	81.64	0.9536
0.3652	-5.77	36.54	0.9750	10.53	79.58	0.9573
0.5963	-7.87	38.10	0.9705	7.58	82.38	0.9544
0.8331	-5.97	38.54	0.9364	8.11	82.71	0.9545
Average		37.53			81.58	
	F - D			Huang		
403.1 K	ord.10^4	±σ.10^5	r^2	ord.10^4	±σ.10^5	r^2
0.1102	5.89	33.96	0.9674	26.53	82.02	0.9534
0.3652	-3.25	34.79	0.9913	14.29	82.77	0.9538
0.5963	-5.28	35.94	0.9414	10.89	86.20	0.9506
0.8331	-3.05	37.07	0.9579	12.12	88.26	0.9487
Average		35.44			84.81	

Table 5. Intercepts (ord), standard deviations (σ), and correlation coefficients (r^2) in the system DTAB-OTAB.

In Huang's methodology, the only negative intercept values correspond to those of the last two weight fractions at the first temperature, but in this methodology the values show incoherence at the different temperatures. The highest errors appear in Huang's method (close to the double) and the correlation coefficients are good and similar in both methodologies (0.94 -0.99 in F-D and 0.94-0.96 in Huang).

In all systems we can point out that the major errors correspond to Huang's methodology, with a good correlation of the lineal regression in this method. This behaviour is similar to that obtained by Benabdelghani et al. (Benabdelghani, et al., 2006) in polymers, and as pointed out by these authors, this could be due to the fact that the experimental points fitted in Huang's method are far from the intercept and its determination is subject to a major error. The enlargement of the values that correspond to the intercept in Huang's methodology leads to the duplication of errors and to results that are less numerically stable in relation to F-D methodology.

It is worth mentioning that besides the intrinsic errors of the chromatographic method in the determination of the parameter χ_{23}/V_2 (obtained from a delicate balance between a ternary system ($\chi_{1(2,3)}$) and two binary systems (χ_{12} and χ_{13})), there would be experimental errors of the method as pointed out by Etxeberria et al. (Etxeberria et al., 2000). One way of reducing the uncertainty in the values of B_{23} is an adequate selection of probe solvents. In this way, we have tried to cover all the possible chemical structures and polarities, avoiding the tests with similar retention among the retentive possibilities in this type of surfactant. We have included hydrocarbons in the probe solutes although they are not recommended for the experimental determination of polymer interaction parameters (Etxeberria et al., 2000).

In Table 2, we have included values of solution enthalpies obtained of the retention time data vs. 1/T, their errors at a level of reliability of 95 % and the differences between the values of solution enthalpies for each of the systems studied. In general terms we can point out that the system DDAB - DODAB presents the slightest differences between ΔH^{o}_{s}, and the minor error among the systems studied. Besides, both stationary phases would show a major apolar character due to the major density of hydrocarbon chain. The other two systems present similar values of solution heat and the highest values in the component with a minor hydrocarbon chain.

We have calculated the average errors in the solution heat of the systems studied, which are also included in Table 2. Thus, we can observe that the system DDAB - DODAB presents a minor average error in the solution heat of 0.81 %, the system DPC - CPB is intermediate with an average value of 1.01 % and the third system DTAB - OTAB presents a major value of 1.21 %. The third system presents the lower values of V^{0}_{g} among the systems we studied and the major experimental errors in this retention parameter. The percentage values of the errors in the system DTAB - OTAB are very different from those calculated in the other systems following the method F - D, and there is a large difference in the errors obtained from a retention time correlation as a function of 1/T.

We have observed that in the results determined in the three systems by Huang's methodology (Huang, 2003a, 2003b) the value of the slope shows that when the withdrawal from the unit is greater, the difference between the intercept values is bigger in both methodologies.

In Table 6 we have included the values obtained from the differences between the intercept values determined by both methods in the systems studied and the slope calculated by Huang's methodology for the first values of temperature.

As it can be seen in Table 6, the highest values of the differences between intercept values are visible when the deviation of the slope from the unit is major. We have plotted the values of the differences between the ordinate values in the method F-D (Farooque, & Deshpande, 1992) and in Huang's methodology (Huang, 2003a, 2003b) as a function of the slope determined by Huang's method. Figure 3 shows that in the systems DDAB-DODAB and DPC-CPB, the dots rest on a straight line, with a major concentration of the values closeness to the unitary slope value. When the slope values are higher, the differences between both methods are greater.

DDAB–DODAB system				
363.1 K	Ord.10^4			
w_2	F-D	Huang	Difference	Huang slope
0.1643	18.64	18.91	-0.27	1.000
0.3281	19.98	19.61	0.37	1.006
0.5037	18.34	17.30	1.04	1.013
0.7325	17.06	15.04	2.02	1.023
DPC-CPB system				
348.1 K	Ord.10^4			
w_2	F-D	Huang	Difference	Huang slope
0.1598	9.41	4.94	4.47	1.037
0.3461	8.48	3.17	5.31	1.045
0.5409	8.66	3.54	5.12	1.044
0.7590	12.07	10.14	1.93	1.018
DTAB-OTAB system				
388.1 K	Ord.10^4			
w_2	F-D	Huang	Difference	Huang slope
0.1102	-4.31	12.45	-16.76	0.8908
0.3652	-3.52	0.45	-13.97	0.8903
0.5963	-15.20	-2.81	-12.39	0.8795
0.8331	-13.89	-1.72	-12.17	0.8563

Table 6. Differences of the values between the intercept values determinate by both methodologies and the slope calculated by Huang's methodology.

The dots corresponding to the system DTAB-OTAB, are in a part of the graph opposite to the values found in the other two systems. In this system, it can be observed the higher differences between the ordinate values, the major errors in both methodologies; the errors in Huang's method are twice as big as those found in F-D, the correlation coefficients are similar and the slopes in Huang's method are minor than the unit.

In general terms, we can point out that the F-D method is numerically more stable than Huang's methodology in these systems. This would be due to the algebraic treatment of the departure expression. The higher values of the correlation coefficient in the regression are the result from the scale expansion in Huang's methodology.

As a conclusion, in these systems the results of the comparison between the methodologies are in agreement with what Benabdelghani et al. (Benabdelghani, et al., 2006) said. We also found that when the value of withdrawal of the slope from the unit in Huang's method is higher, the values are less reliable and there are higher errors than in the method of F-D.

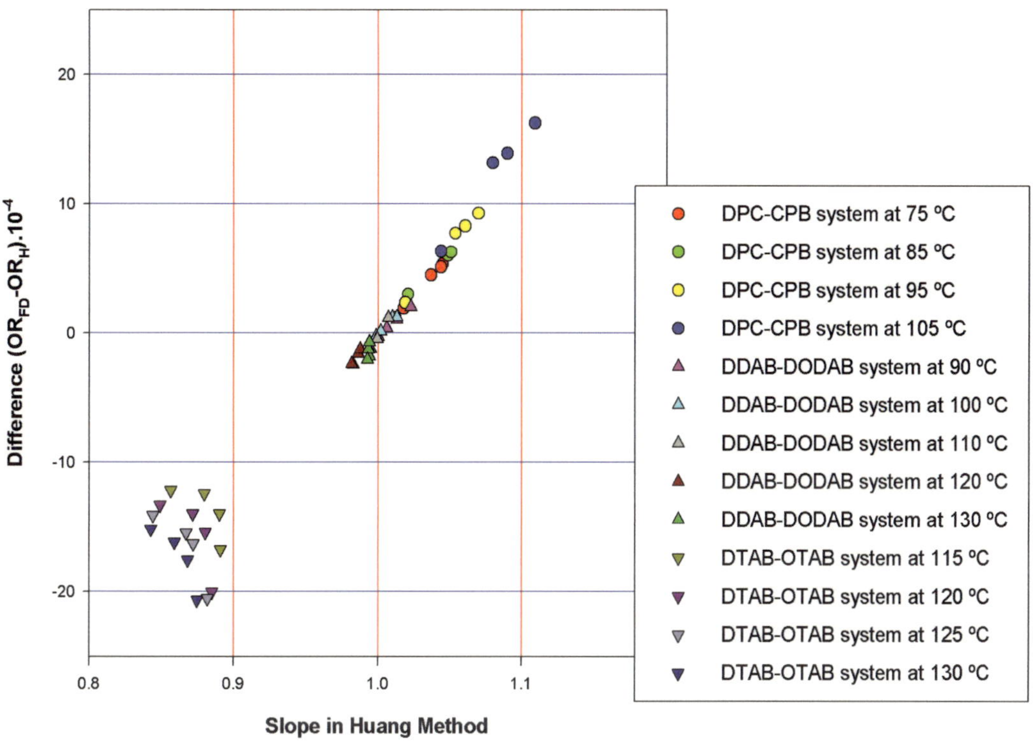

Fig. 3. Differences between the ordinate values in the method F-D and in Huang's methodology as a function of the slope determined by Huang's method.

5. Conclusions

The studied systems were different mixtures of anhydrous cationic surfactants, deposited on solid support, that has been used as stationary phases in IGC and the results were analyzed in term of mixtures miscibility.

The determination of the thermodynamic miscibility was realized by the values of surfactant-surfactant interaction parameters obtained by the same method used in polymeric materials.

The comparative use of two phenomenological methodologies allowed us to calculate the surfactant-surfactant interaction parameter in anhydrous cationic surfactant mixtures; the obtained results were similar to those obtained by Benabdelghaie et al. (Benabdelghani, et al., 2006) in polymeric materials.

We also found that when the value of withdrawal of the slope from the unit in Huang's method is higher, the values are less reliable and there are major errors than in the method of F-D.

6. Acknowledgments

This work was sponsored by CIUNT (Consejo de Investigaciones de la Universidad Nacional de Tucumán) and partially by UNS. E.L.A. is a member of CONICET (Consejo Nacional de Investigaciones Científicas y Técnicas de la Republica Argentina).

7. References

Abe, A. & Flory, P.J., (1978). Statistical thermodynamics of mixtures of rodlike particles. 2. Ternary systems. *Macromolecules*, 11, 1122-1126

Al-Saigh, Z. Y., & Munk, P., (1984). Study of polymer-polymer interaction coefficients in polymer blends using Inverse Gas Chromatography. *Macromolecules*, 17, 803-809

Akisada, H. , Kuwahara, J., Koga, A., Motoyama, H. & Kaneda, H., (2007). Unusual behavior of CMC for binary mixtures of alkyltrimethylammonium bromides: Dependence on chain length difference. *J. Colloid Interface Sci.*, 315, 678-684

Bardavid, S. M., Schulz, P. C. & Arancibia, E. L., (2003). Solubility parameter determination of cationic surfactant by Inverse GC. *Chromatographia*, 57, 529-532

Bardavid, S. M., Schulz, P. C. & Arancibia, E. L., (2007). IGC studies of binary cationic surfactant mixtures. *J. Colloids Interface Sci.*, 316, 114-117

Bardavid, S. M., Schulz, P. C. & Arancibia, E. L., (2010). Miscibility of anhydrous cationic surfactant mixtures. *J. Mol. Liquids*, 156, 165-170

Bardavid, S.M., Schulz, P.C. & Arancibia, E.L., (2011). Interaction parameter of anhydrous cationic surfactant mixtures by IGC. Acepted in the *J. Solution Chemistry* (2011)

Benabdelghani, Z., Etxeberria, A., Djadoun, S., Uruin, J. J. & Uriarte, C., (2006). The phase behaviour of poly(styrene-co-methacrylic acid)/poly(2,6-dimethyl-1,4-phenylene oxide) by Inverse Gas Chromatography. *J. Chromatog. A*, 1127, 237-245

Bhat, M. A., Dar, A. A., Amin, A., Rashid, P. I. & Rather, G. M., (2007). Temperature dependence of transport and equilibrium properties of alkylpyridinium surfactants in aqueous solutions. *J. Chem. Thermodyn.*, 39, 1500–1507

Blackmore, E. S. & Tiddy, G. J. T. , (1990). Optical microscopy, multinuclear NMR (^{2}H, ^{14}N and ^{35}Cl) and X-ray studies of dodecyl-and hexadecyl-trimethylammonium chloride/water mesophases. *Liquid Crystals*, 8, 131-151

Chow, L. C. & Martire, D. E., (1971). Thermodynamics of solutions with liquid crystal solvents. III. Molecular interpretation of solubility in nematogenic solvents. *J. Phys. Chem.*, 75, 2005-2015

Conder, J. R. & Young, C. L., (1978) *Physicochemical Measurement by Gas Chromatography*, Wiley, New York

Deshpande, D. D., Patterson, D., Screiber, H. P. & Su, C. S., (1974). Thermodynamic interactions in polymer Systems by Gas-Liquid-Chromatography. IV.

Interactions between components in a mixed stationary phase. *Macromolecules*, 7, 530- 535.

DiPaola-Baranyi, G. & Degre, P., (1981). Thermodynamic characterization of polystyrene-poly(butyl methacrylate) blends. *Macromolecules*, 14, 1456-1460.

Dreisbach, R. R., (1955) *Advances Chemistry Series*, publishing by A.C.S. Du, Q., Chen, W. & Munk P., (1999). Inverse Gas Chromatography. 8. Apparent probe dependence of χ_{23}' for a poly(vinil chloride)-poly(tetramethylene glycol) blend. *Macromolecules*, 32, 1514-1518

El-Hibri, M. J., Cheng, W., & Munk, P., (1988). Inverse Gas chromatography. 6. Thermodynamics of poly(ε-caprolactone)-polyepichlorohydrin blends. *Macromolecules*, 21, 3458-3463

Etxabarren, C., Iriarte, M., Uriarte, C., Etxeberria, A. & Iruin, J. J., (2002). Polymer-solvent interaction parameters in polymer solutions at high polymer concentrations. *J. Chromatogr. A*, 969, 245-254

Etxeberria, A., Uriarte, C., Fernandez-Berridi, M.J. & Uruin, J.J., (1994). Probing polymer-polymer interaction parameters in miscible blends by Inverse Gas Chromatography: Solvent effects. *Macromolecules*, 27, 1245-1248

Etxeberria, A., Iriarte, M., Uriarte, C., & Iruin, J. J., (1995). Lattice Fluid Theory and Inverse Gas Chromatography in the analysis of polymer-polymer interactions. *Macromolecules*, 28, 7188-7195

Etxeberria, A., Etxabarren, C. & Iruin, J. J., (2000). Comparison between static (sorption) and dynamic (IGC) methods in the determination of interaction parameters in polymer/polymer blends. *Macromolecules*, 33, 9115-9121

Farooque, A. M. & Deshpande, D. D., (1992). Studies of polystyrene-polybutadiene blend system by inverse gas chromatography. *Polymer*, 33, 5005-5018.

Feitosa, E., Alves, F. R., Niemiec, A., Real Oliveira, M. E.C. D., Castanheira, E. M. S. & Baptista, A. L. F., (2006). Cationic liposomes in mixed didodecyldimethylammonium bromide and dioctadecyldimethylammonium bromide aqueous dispersions studied by Differential Scanning Calorimetry, Nile Red Fluorescence, and Turbidity. *Langmuir*, 22, 3579-3585

Flory, P. J., (1953) Principles *of Polymer Chemistry*, Cornell University Press, Ithaca (NY)

Flory, P. J., (1956). Statistical thermodynamics of semi-flexible chain molecules. Proc. R. Soc., Lond. A, 234, 60-73

Flory, P. J. & Abe, A., (1978). Statistical thermodynamics of mixtures of rodlike particles. 1. Theory for polydisperse systems. *Macromolecules*, 11, 1119-1122

Flory, P. J., (1984). Molecular theory of liquid crystals. *Liquid Crystal Polymers I. Advances in Polymer Science*, 59, 1-36

Fuerstenau, D. W. & Jia, R., (2004). The adsorption of alkylpyridinium chlorides and their effect on the interfacial behavior of quartz. *Colloids and Surfaces A: Physicochem. Eng. Aspects*, 250, 223–231

Holland, P. M. & Rubingh, D. N., (1990) In *Cationic Surfactants*; Holland, P. M. & Rubingh, D. N. , Eds.; Surfactant Science Series; Marcel Dekker, New York, *Vol.* 37

Huang, J. C., (2003a). Analysis of the thermodynamic compatibility of poly(vinyl chloride) and nitrile rubbers from Inverse Gas Chromatography. *J. Appl. Polym. Sci.*, 89, 1242-1249

Huang, J. C., (2003b). Determination of polymer-polymer interaction parameters using Inverse Gas Chromatography. *J. Appl. Polym. Sci.*, 90, 671-680

Huang, J. C., Coca, J. & Langer, S. H., (2007). Liquid crystal solutions at infinite dilution: Solute phase transfer free energy and solubility parameter variations at phase conversion temperatures. *Fluid Phase Equilib.*, 253, 42-47

Janini, G. M., Johnston, K. & Zielinski, W. L. Jr., (1975). Use of a nematic liquid crystal for gas- liquid chromatographic separation of polyaromatic hydrocarbons. *Anal. Chem.*, 47, 670- 674

Janini, G. M., Manning, W. B., Zielinski, W. L. Jr.& Muschik, M., (1980). Gas-liquid chromatographic separation of bile acids and steroids on a nematic liquid crystal. *J. Chromatogr.*, 193, 444-450

Laughlin, R. G., (1990). *Cationic Surfactants,* Physical Chemistry, Ed. Rubingh, D. N. & Holland, P. M., Marcel Dekker, New York

Laughlin, R., (1994). *The aqueous phase behaviour of surfactants,* Academic Press Inc., San Diego, CA

Lezcano, E. G., Prolongo, M. G. & Salom Coil, C., (1995). Characterization of the interactions in the poly(4-hydroxystyrene)/poly(s-caprolactone) system by Inverse Gas Chromatography. *Polymer*, 36, 565-573

Marciniak, W. & Witkiewicz, Z., (1981). Effect of the amount of liquid crystal and type of support on some properties of the liquid crystalline stationary phase-support system. *J. Chromatogr.* , 207, 333-343

Marquartd, D. W., (1963). An Algorithm for Least-Squares Estimation of Nonlinear Parameters. *SIAM Journal on Applied Mathematics* 11 (2) 431–441

Martire, D. E., Blasco, P. A., Carone, P. E., Chow, L. C.& Vicini, H., (1968). Thermodynamics of solutions with liquid-crystal solvents. I. A Gas-Liquid Chromatographic study of cholesteryl myristate. *J. Phys. Chem.*, 72, 3489-3495

Nastasovic, A. B. & Onjia, A. E., (2008). Determination of glass temperature of polymers by Inverse Gas Chromatography. *J. Chromatog. A*, 1195, 1-15

Patterson, D., Tewari, Y.B. , Schreiber, H.B. & Guillet, J.E., (1971). Application of Gas-Liquid Chromatography to the thermodynamics of Polymer Solutions. *Macromolecules* , 4, 356-359

Price, G. J., Hickling, S. J. & Shillcock, I. M., (2002). Applications of Inverse gas Chromatography in the study of liquid crystalline stationary phases. *J. Chormatogr. A*, 969, 193-205

Prolongo, M. G., Masegosa, R. M. & Horta, A., (1989). Polymer-polymer interaction parameter in the presence of a solvent. *Macromolecules*, 22 4346-4351

Proverbio, Z. E., Bardavid, S. M., Arancibia, E. L. & Schulz, P. C., (2003). Hydrophile-lipophile balance and solubility parameter of cationic surfactants. *Colloids and Surf. A, Physicochem. Eng. Aspects*, 214, 167-171

Rayss, J., Witkiewicz, Z. , Waksmundzki, A. & Dabrowski, R., (1980). Effect of the support surface on the structure of the film of liquid crystalline stationary phase. *J. Chromatgr. A.*, 188, 107-113

Reid, R. C., Prausnitz, J. M. & Poling, B. E., (1986). *The properties of Gases and Liquids*, 4th Ed., McGraw-Hill, New York

Riddick, A. R., Bunger, W. B. & Sakano, T. K., (1986). *Organic Solvents*. Techniques of Chemistry, 4th Ed., Wiley-Interscience, New York

Romannsky, M. & Guillet, J. E., (1994). The use of Inverse Gas Chromatography to study liquid crystalline polymers. *Polymer*, 35, 584-589

Romannsky, M. , Smith, P. F., Guillet, J. E. & Griffin, A. C., (1994). Solvent interactions with an insoluble liquid-crystalline polyester. *Macromolecules*, 27, 6297-6300

Schaefer, C. R. de, Ruiz Holgado, M. E. F. de & Arancibia, E. L., (2008). Effective solubility parameters of sucrose monoester surfactants obtained by Inverse Gas Chromatography. *Fluid Phase Equilib.*, 272, 53-59

Shi, Z. H. & Screiber, H. P., (1991). On the application of Inverse Gas Chromatography to interactions in mixed stationary phases. *Macromolecules*, 24, 3522-3527

Shillcock, L. M. & Price, G. J. , (2003). Inverse Gas Chromatography study of poly(dimethyl siloxane)-liquid crystal mixtures. *Polymer*, 44, 1027-1034

Shillcock, L. M. & Price, G. J. , (2004). Interactions of solvents with low molar mass and side chain polymer Liquid Crystals measured by Inverse Gas Chromatography. *J. Phys. Chem. B*, 108, 16405- 16414

Schulz, P. C. , Puig, J. E. , Barreiro, G. & Torres, L. A. (1994) . Thermal transitions in surfactant- based lyotropic liquid crystals. *Thermochimica Acta* 231, 239-256

Schulz, P.C., Abrameto, M., Puig, J.E., Soltero-Martínez, F.A. & González-Alvarez, A., (1996). Phase Behavior of the Systems n-Decanephosphonic Acid - Water and n-Dodecanephosphonic Acid- Water. *Langmuir*, 12, 3082-3088

Schulz, P. C. , Rodriguez, J. L., Soltero-Martinez, A. Puig, J. E. & Proverbio, Z. E., (1998). Phase behaviour of the dioctadecyldimethylammonium bromide-water system. *J. Thermal Anal.*, 51, 49-62

Schulz, P. C., Rodríguez, J. L., Minardi, R. M., Sierra, M. B. & Morini, M. A., (2006). Are the mixtures of homologous surfactants ideal? *J. Colloid & Interface Sci.*, 303, 264-271

Tompa, H., (1956). *Polymer Solutions*, Butterworths, London Varade, D., Aramaki, K. & Stubenrauch, C., (2008). Phase diagrams of water– alkyltrimethylammonium bromide systems. *Colloids Surf. A: Physicochem. Eng. Aspects*, 315, 205-209

Voelkel, A., Strzemiecka, B., Adamska, K. & Milczewska, K., (2009). Inverse Gas Chromatography as a source of physicochemical data. *J. Chromatogr. A*, 1216, 1551-1566

Zhao, L. & Choi, P., (2001). Determination of solvent-independent polymer-polymer interaction parameter by an improved Inverse Gas Chromatographic approach. *Polymer*, 42, 1075-1081

Zhikuan, C. & Walsh, D. J., (1983). Inverse Gas Chromatography for the study of one phase and two phase polymer mixtures. *Eur. Polym. J.*, 19, 519- 524

Zhou, Y. W. , Jaroniec, M., Hann, G. L. & Gilpin, R. K., (1994). Gas Chromatographic and Infrared studies of 4'-cyano-4-biphenyl 4-(4-pentenyloxy)benzoate coated on porous silica. *Anal. Chem.*, 66, 1454- 1458

3

Applications of Chromatography Hyphenated Techniques in the Field of Lignin Pyrolysis

Shubin Wu, Gaojin Lv and Rui Lou
South China University of Technology, China

1. Introduction

Due to the urgency of the current world energy supply-and-demand situation, the need for clean sources of energy is receiving an increasing attention. In the framework of a future sustainable development, biomass is one of the most often considered sources of renewable energy (Bridgewater et al., 1999). There are many ways of converting biomass into useful products and energy, such as direct combustion processes, thermochemical processes, biochemical processes, and agrochemical processes etc. Of these, pyrolysis forms the focus of this study (Bridgewater, 2004; Mohan et al., 2006). The pyrolysis of lignocellulose is very complex, primarily due to the inherent complexity of the substrate, which changes continuously both chemically and structurally throughout the decomposition process (Hosoya et al., 2007; Lv et al., 2010a).

The chemical structure and major organic components in biomass are extremely important in biomass pyrolysis processes. Knowledge of the pyrolysis characteristics of the three main components is the basis and thus essentially important for a better understanding of biomass thermal chemical conversion. Lignin is one of the main components of woody biomass, and the worldwide production of technical lignins as a by-product from chemical pulping processes stands at approximately 50 million t/yr (Harumi et al., 2010). However, it is merely used as fuel to recover energy in conventional pulping industry. Only recently, with the upcoming focus on biorefineries, lignin has gained new interest as chemical resources.

Analytical pyrolysis is a well-known technique to analyse lignin pyrolysis and various authors published different analytical methods to determine decomposition characteristics of lignin. Many researchers presented that the pyrolysis of lignin primarily occurred in a wide temperature range (200-600°C) by means of thermogravimetric analysis method (Domínguez et al., 2008; Lv et al., 2010b). Some researchers (Liu et al., 2008; Wang et al., 2009) also compared the pyrolysis behavior of lignin from different tree species using thermogravimetry–Fourier transform infrared spectroscopy (TG–FTIR). Pyrolysis–gas chromatography/mass spectrometry (Py-GC/MS), which is an advanced pyrolysis methods combined with hyphenated separation and detection systems (i.e. GC-MS), is often used for studying degradation mechanisms of lignin because of its strong ability to identify the pyrolysis products (Atika et al., 2007; Windt et al., 2009).

On lignin pyrolysis, their product analysis, and further deriving the cracking mechanism, many researchers have done considerable works. For example, Baumlin et al. (2006) have reported the results of experiments performed on the flash pyrolysis of two types of lignins, i.e. kraft lignin and organocell lignin, to produce hydrogen. Nowakowskia et al. (2010) presented an international study of fast pyrolysis of lignin. Jegers and Klein (1985) had reported that various catechol (1,2-dihydroxybenzene), o-cresol (2-methylphenol) and other phenols products go along with the formation of guaiacols during the course of lignin pyrolysis, the yields of those products are different corresponding to the different kinds of lignin and pyrolysis conditions. According to Britt et al. (1995), lignin pyrolysis occurred mainly by a free-radical reaction mechanism. The relative distribution of products is dependent on pyrolysis conditions, such as sources of raw materials, pyrolysis temperature, heating rate, pyrolysis atmosphere, and catalyst etc. (Garcia et al., 2008).

To sum up, there have already emerged lots of studies about lignin pyrolysis, and their focus and concerns were also varied. But so far, to our knowledge, limited information is available in the literature concerning the product generation and distribution regularities of lignin pyrolysis under the influence of parameters like temperatures and catalysts. Therefore, the objectives of this work were to attempt to carry out fast pyrolysis of several lignin samples (one enzymatic/mild acidolysis lignin and two technical lignins were used) and analyse the products by Py-GC/MS, in order to firstly establish the potential of this method for lignin processing and secondly to investigate the effects of temperature and catalysts on lignin pyrolysis.

2. Hyphenated techniques

In the case of samples originating in the real world, each of the techniques has a place, and often several must be used in order to obtain a complete overview of the nature of the sample.

The use of multiple techniques and instruments, which allow more than one analysis to be performed on the same sample at the same time, provide powerful methods for analyzing complex samples (Kealey & Haines, 2002). If the instruments are combined so that the analyses are done essentially at the same time, this is called a simultaneous approach and is often written with a hyphen, so that they may be referred to as hyphenated techniques, for example gas chromatography-mass spectrometry (GC-MS) and gas chromatography-infrared spectrometry (GC-IR). By using many techniques in combinations, the advantages to the analyst in the additional information, time saving and sample throughput are considerable.

2.1 Gas Chromatography-Mass Spectrometry (GC-MS)

The use of chromatographic techniques to separate mixtures is one of the most important analytical tools. The separated components may then be identified by other techniques. Mass spectrometry is the most important of these.

2.1.1 Separation

Gas chromatography, is a common type of chromatography used in analytic chemistry for separating and analyzing compounds that can be vaporized without decomposition. Typical uses of GC include testing the purity of a particular substance, or separating the different components of a complex mixture, such as bio oil.

When used to prepare pure compounds from a mixture, GC can separate the volatile components of mixtures by differential migration through a column containing a liquid or solid stationary phase (Fu, 2008). Solutes are transported through the column by a gaseous mobile phase and are detected as they are eluted. Solutes are generally eluted in order of increasing boiling point, except where there are specific interactions with the stationary phase. An elevated temperature, usually in the range 50-350°C, is normally employed to ensure that the solutes have adequate volatility and are therefore eluted reasonably quickly.

2.1.2 Identification

Mass spectrometry (MS) is an analytical technique in which gaseous ions formed from the molecules or atoms of a sample are separated in space or time and detected according to their mass-to-charge ratio, m/z (Sparkman, 2000). It is usually used for determining masses of particles, for determining the elemental composition of a sample or molecule, and for elucidating the chemical structures of molecules, such as phenols, aldehydes, and other chemical compounds.

The MS principle consists of ionizing chemical compounds to generate charged molecules or molecule fragments and measuring their mass-to-charge ratios. The numbers of ions of each mass detected constitute a mass spectrum. The spectrum provides structural information and often an accurate relative molecular mass from which an unknown compound can be identified or a structure confirmed.

2.1.3 Combination

Gas chromatography-mass spectrometry (GC-MS) is a common combined technique, comprising a gas chromatograph (GC) coupled to a mass spectrometer (MS), by which complex mixtures of chemicals may be separated, identified and quantified. A schematic diagram of a GC-MS is shown in Fig. 1. In this technique, a gas chromatograph is used to separate different compounds. This stream of separated compounds is fed online into the ion source, a metallic filament to which voltage is applied. This filament emits electrons which ionize the compounds. The ions can then further fragment, yielding predictable patterns. Intact ions and fragments pass into the mass spectrometer's analyzer and are eventually detected (Adams, 2007; Lee & Eugene, 2004).

This makes it becoming an ideal tool of choice for the analysis of the hundreds of relatively low molecular weight compounds found in biomass pyrolysis liquid products (bio oil). In order to make a compound be analysed by GC-MS, it must be sufficiently volatile and thermally stable. In addition, functionalised compounds may require chemical modification (derivatization) prior to analysis, to eliminate undesirable adsorption effects that would otherwise affect the quality of the data obtained (Wu, 2005). Bio oil samples are usually needed to be solvent extracted, and dehydrated before GC-MS analysis.

The prepared sample solution is injected into the GC inlet where it is vaporized and swept into a chromatographic column by the carrier gas. The sample flows through the column and the compounds comprising the mixture of interest are separated by virtue of their

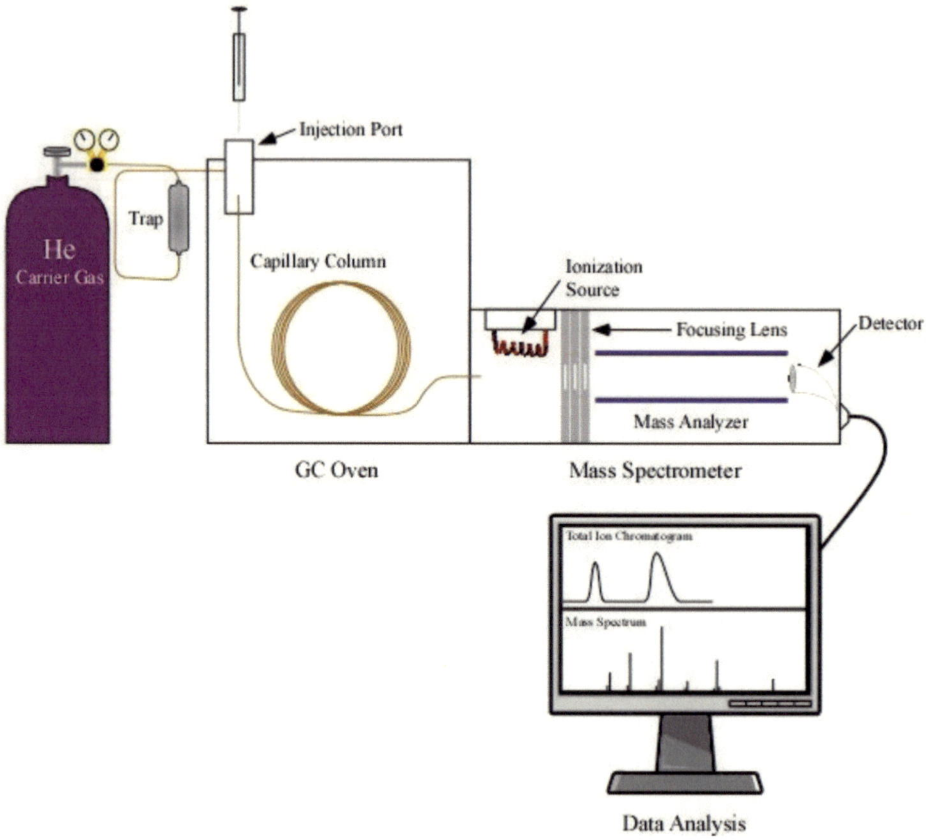

Fig. 1. Schematic of a GC-MS system

relative interaction with the coating of the column (stationary phase) and the carrier gas (mobile phase). The latter part of the column passes through a heated transfer line and ends at the entrance to ion source (Fig. 1) where compounds eluting from the column are converted to ions. Then the ions are separated in a mass analyser (filter). After that, the ions enter a detector the output from which is amplified to boost the signal. The detector sends information to a computer that records all of the data produced, converts the electrical impulses into visual displays and hard copy displays. In addition, the computer also controls the operation of the GC-MS system.

2.2 Pyrolysis-Gas Chromatography/Mass Spectrometry (Py-GC/MS)

Pyrolysis-gas chromatography/mass spectrometry (Py-GC/MS), which extends the combination to include three distinct techniques, is an instrumental method that enables a reproducible characterisation of the intractable and involatile macromolecular complexes found in virtually all materials in the natural environment (Bull, 2005). It differs from GC-MS in the type of sample analysed and the method by which it is introduced to the GC-MS system. Instead of the direct injection of a highly refined organic solution, a few amount (usually<mg) of the original natural material (e.g. soil, sediment, biomass etc.) is analysed directly (Jin et al., 2009).

When analyzed, the samples are first inserted into a quartz chamber in a pyrolysis unit (Fig. 2) that is then heated resistively in an oxygen free environment at a pre-set temperature for a number of seconds (e.g. 600°C for 10s). This results in a heat mediated cleavage of chemical bonds within the macromolecular structures of interest producing a suite of low molecular weight chemical moieties, which is indicative of specific types of macromolecule (e.g. lignin, cellulose, hemicelluloses etc.). This mixture of compounds is then swept into the analytical column of the GC and GC-MS proceeds as normal.

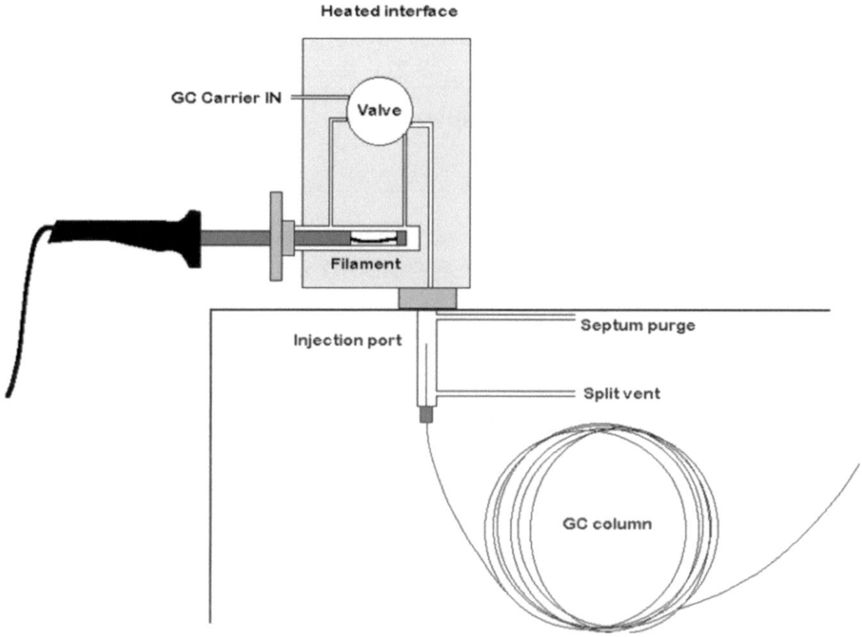

Fig. 2. Schematic of a Pyroprobe combined with GC-MS system

Because of its high sensitivity, a rapid analysis time, and less sample pretreatments, the analytical technique of Py–GC/MS is widely used to research chemical structure and pyrolysis characteristics of biomass and its three main components (i.e. lignin, cellulose, and hemicellulose), to examine reaction products of biomass thermal degradation (Meier & Faix, 1999), and to investigate fast pyrolysis of biomass and on-line analysis of the pyrolysis vapors (Fahmi et al., 2007; Nowakowski & Jones, 2008)

3. Experimental

3.1 Materials

The enzymatic/mild acidolysis lignin (EMAL) was isolated from Moso bamboo by means of enzymatic/mild acidolysis method previously described (Wu & Argyropoulos, 2003; Lou & Wu, 2011), and the so called enzymatic/mild acidolysis lignin was obtained.

Technical alkali lignin (AL) and acid hydrolysis lignin (AHL) were isolated from bagasse black liquor and bagasse respectively, according to our previous reports (Wu et al., 2008; Tan, 2009).

Elemental analysis of three types of lignin was implemented in a Vario EL elemental analyzer and an ICP inductively coupled plasma emission spectrometer. Table 1 lists the results of the elemental analysis, from which the O content can be calculated by difference.

	Organic Elements / wt%				Inorganic elements / ppm								
	C	H	N	S	Al	Ca	Na	Mg	Mn	K	Zn	Cu	Fe
EMAL	58.74	5.72	2.58	0.06	5.73	1.26	80.19	0.37	-	12.50	14.45	3.79	109.10
AL	62.20	7.37	0.07	0.65	-	381.79	226.70	26.13	3.56	806.27	9.87	32.81	89.81
AHL	49.64	5.78	2.52	0.17	8.55	189.68	47.22	16.95	1.23	19.13	0.46	26.70	116.37

Table 1. Elemental analysis of the lignin samples

3.2 Analytical pyrolysis

Fast pyrolysis of prepared samples was carried out in a Py-GC/MS system, which includes a JHP-3 model Curie-point pyrolyzer (CDS5200, USA) and a Shimadzu QP2010 Plus gas chromatograph-mass spectrometer (Japan). The pyrolyzer consisted of an inductive heated coil to heat the samples and was capable of maintaining up to 1200°C temperature with a heating rate of approximately 10°C· ms^{-1} from room temperature to the terminal temperature, with helium as both purge gas and carrier gas.

On the basis of the thermal behaviors of EMAL (Lou & Wu, 2008), the pyrolysis temperature of EMAL samples was set at 320°C, 400°C, 600°C, and 800°C, respectively. Approximately 0.1mg of each sample was pyrolyzed. Pyrolysis reactions were carried out with an event time of 10s, and the obtained pyrolysis products were then analyzed by GC-MS.

The pyrolysis products were separated in a DB-5MS (Agilent Technologies, USA) capillary column (30m×0.25mm×0.25μm). The split ratio of 70/1 and linear velocity of 40.0cm · s^{-1} was used. The GC oven was heated from 50°C to 250°C at a heating rate of 10°C/min, and then maintained for another 2min. The injector temperature was 250°C. The mass spectrometer was operated in the EI mode using 70eV of electron energy. The mass range m/z 45-500 was scanned. Identification of the pyrolysis compounds was achieved by comparison of their mass fragment with Perkin Elmer NIST 05 mass spectral library. For qualitative and quantitative analysis of the pyrolysis products, a more detailed explanation based on a practical example will be given.

3.3 Catalytic pyrolysis

In order to study the effect of catalysts on EMAL pyrolysis, sodium chloride (NaCl) of metal salt and permutite of zeolite were selected to serve as the catalyst. The catalysts were previously subject to high temperature treatment under N_2 atmosphere at 800°C for 6h to ensure the decomposition reaction of catalysts did not happen during lignin pyrolyzing. The additive amount of each catalyst was set to be 5%, 10%, and 20% based on weight, and the catalyst and lignin were ground and mixed together evenly before used for experiment. As lignin pyrolysis reaction was complete at 800°C, so the catalytic pyrolysis temperature of EMAL was set at 800°C to better investigate the impact of the additive catalysts on products.

4. Results and discussion

4.1 Identification of pyrolysis products

According to the previous introduction of GC-MS, both qualitative and quantitative analysis of the pyrolysis compounds can be achieved with GC-MS easily. For example, Fig. 3 shows TIC chromatogram of Bamboo lignin pyrolysis at 600°C, in which each peak shown a compound produced during lignin pyrolysis process. An effective and efficient way to qualitative identify these peaks is to compare its experimental mass spectrum against a library of computerized mass spectra (Mistrik, 2004).

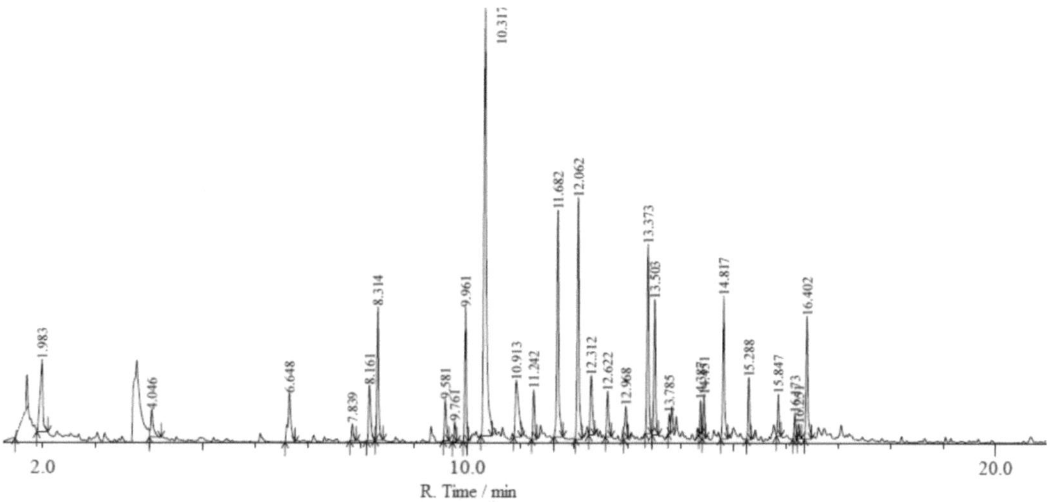

Fig. 3. Total ion chromatograms of Bamboo EMAL pyrolyzed at 600°C

In our experiment analysis, identification of the pyrolysis compounds was achieved by comparison of their mass fragment with Perkin Elmer NIST 05 mass spectral library. Fig. 4 shows examples of identified peaks of some major pyrolysis products by searching computerized spectra library. In addition, by means of spectral libraries, more information about identified compounds can be obtained, such as compound name, molecular formula, structural formula, and molecular weight, etc.

Quantitative analysis of lignin pyrolysis products can be obtained in terms of peak areas or peak heights, both of which are known as semi-quantitative method. Although it can not be accurately determined the content of a certain compound without standard samples, it does can be used to compare the relative content of each or each kinds of compounds, and get the increase or decrease tendencies of a certain kinds of compound with pyrolysis parameters by their peak area percentage. This is why many researchers have been studying biomass pyrolysis by using Py-GC/MS. Of course, for precise quantification of certain components of bio-oil derived from pilot laboratory equipment or factory, it is recommended to use internal or external standards, or by standard addition or internal normalization. However, at present, in order to study the trends or regularities of lignin pyrolysis products with the experimental conditions, quantitative information obtained from integrated peak areas are the most reliable and convenient.

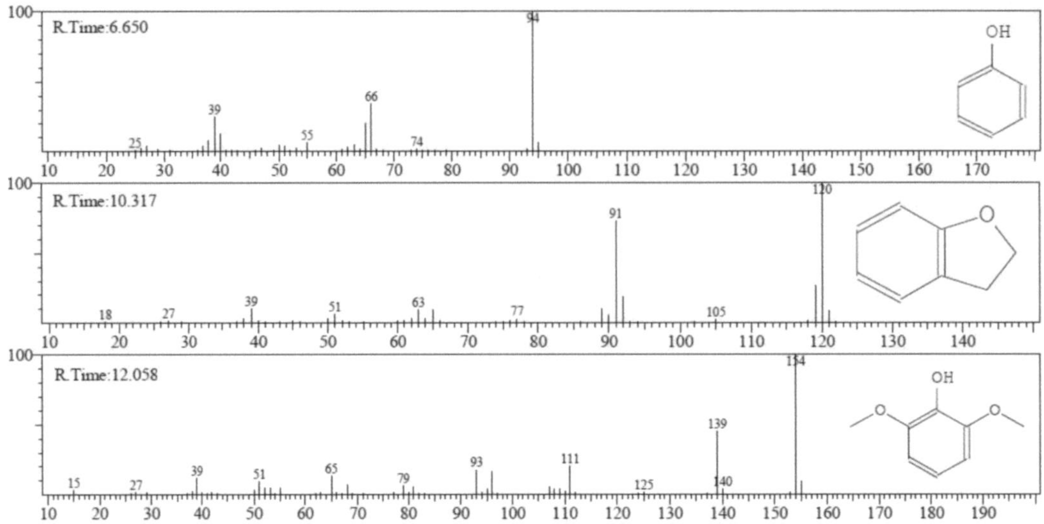

Fig. 4. Some examples of mass spectra identified by searching libraries

4.2 Effect of temperature on EMAL pyrolysis

Based on the analysis method described earlier, quantitative information of bamboo EMAL pyrolysis products at different temperatures are presented in Table 2. As can be seen from Table 2, The major compounds derived from p-hydroxyphenylpropanoid, guaiacylpropanoid, and syringylpropanoid of lignin units during pyrolytic reactions, were mainly classified as the heterocycles (2,3-dihydrobenzofuran), phenols, and a small amount of acetic acid. The yield of phenolic compounds increased with an increase of pyrolysis temperature, and the highest fraction of phenols was 56.43% at 600°C.

Among these pyrolysis products, the small molecule compounds of vanillin and acetic acid generated as a result of the bond cracking of interlinkage C_α-C_β of lignin phenylpropane, and the breakage of C_α-C_β can induce the production of carboxylic acid and carbon dioxide (Yang et al., 2010). Bond breakage of the side chains of lignin structural units can lead to generate degradation products with the new hydroxyl and carbonyl groups. Thus, with the contents of the hydroxyl and carbonyl groups increasing, the side chains of aromatic compounds connected to α-carbonyl, α-carboxyl or ester groups appeared (Lou et al., 2010a).

In all the identified products, 2, 3-dihydrobenzofuran (*DHBF*) accounted for the largest quantity, in addition, other compounds such as ethenylguaiacol, 2, 6-di-tert-butyl-p-cresol (*DTBC*), 3, 5-dimethoxyacetophenone (*DMAP*), methoxyeugenol etc. also account for considerable amount. Some of these selected compounds with higher yields during lignin pyrolysis and their chemical structures are shown in Fig. 5, and their yield distributions varying with pyrolysis temperature are presented in Fig. 6. The yields of obtained compounds possessing the syringyl unit structure (methoxyeugenol, syringol, and syringaldehyde) and 2, 3-dihydrobenzofuran (*DHBF*) are shown in Fig. 6(a) and the yield of compounds possessing guaiacyl unit structure (ethenylguaiacol, vanillin, E-isoeugenol, and sinapylaldehyde) are shown in Fig. 6(b).

Compound class	Compounds	Yield, Area percent (%) [a]			
		320°C	400°C	600°C	800°C
p-Hydroxyphenols	Phenol	- [b]	-	1.61	2.99
	o-Cresol	-	-	-	1.68
	p-Cresol	-	-	1.01	3.58
	2,6-Di-tert-butyl-p-cresol (*DTBC*)	6.91	5.89	2.10	1.67
	2,4-xylenol	-	-	-	0.88
	p-Ethylphenol	-	-	0.74	1.59
	2-Allylphenol	-	-	-	1.84
Guaiacols	Guaiacol	-	-	3.78	1.08
	p-Methylguaiacol	-	-	1.45	1.72
	Ethenylguaiacol	8.18	9.22	7.23	3.16
	Vanillin	-	2.76	2.64	2.51
	E-isoeugenol	-	1.99	2.66	2.00
	Sinapylaldehyde	-	1.50	1.96	2.01
	Coniferylalcohol	-	5.34	9.49	-
Syringols	Syringol	-	3.57	9.69	4.53
	Methylsyringol	-	-	2.98	3.01
	Syringaldehyde	-	2.36	2.42	1.45
	Methoxyeugenol	4.65	5.67	5.06	2.39
	Acetosyringone	-	-	-	1.23
Heterocycles	2,3-Dihydrobenzofuran (*DHBF*)	66.26	56.62	36.05	19.15
Others	Acetic acid	-	-	-	0.97
	3,7-Dimethylnonane	-	-	-	-
	3,5-Dimethoxyacetophenone (*DMAP*)	4.09	5.10	6.63	2.20
	m-Phthalic acid	-	-	-	9.69
	p-p'-Isopropylidenebisphenol (*IPBP*)	-	4.46	5.72	7.89
	Allylphthalate	0.89	0.97	1.37	2.69
	Tetracosane	-	-	-	3.65
	2-Phenylbutyrophenone	-	-	-	2.61
	Dipropylene giycol dibenzote	-	-	-	1.63
	Dotriacontane	-	-	-	4.89

[a] based on the integrated areas; [b] not detected.

Table 2. Products identification from bamboo EMAL pyrolysis

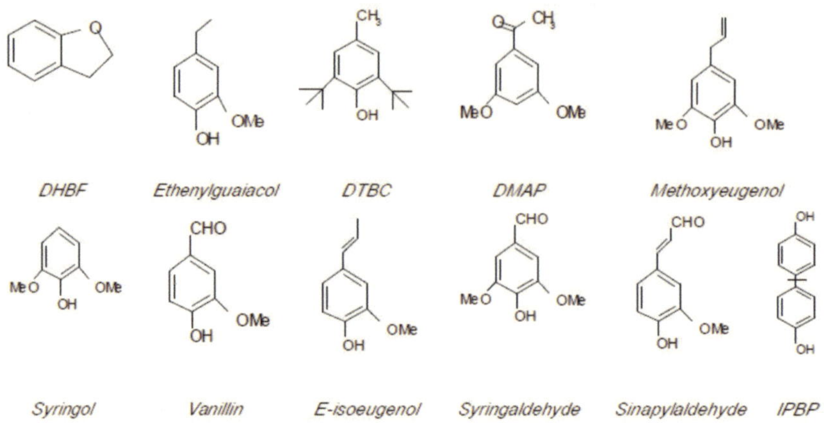

Fig. 5. Products with higher yields in EMAL pyrolysis

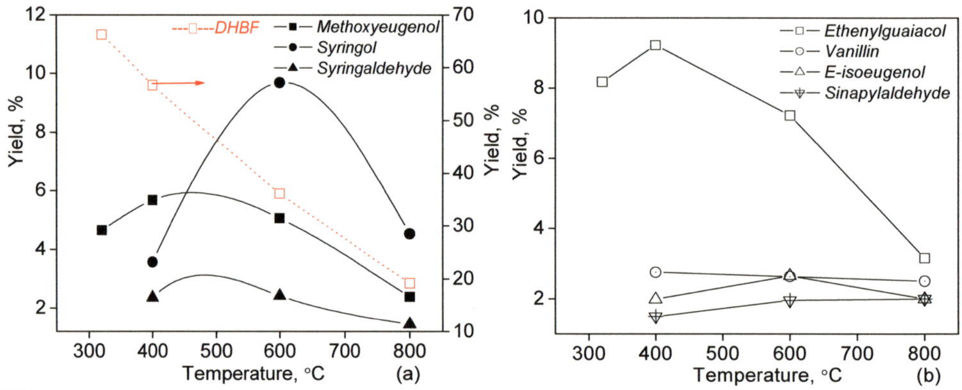

Fig. 6. Product Yields distribution varied with pyrolysis temperature

It can be seen from Fig. 6 that almost all of the compounds derived from EMAL pyrolysis appeared at 400°C, except for *DHBF*, methoxyeugenol, and ethenylguaiacol emerged earlier at about 300°C. The formation fraction of methoxyeugenol, syringaldehyde, ethenylguaiacol, and vanillin reached a maximum at 400°C, then, the yields of these compounds decreased with the increment of temperature, which may be because the secondary decomposition took place at high temperatures. The highest yield of 2, 3-dihydrobenzofuran (*DHBF*) was 66.26% at 320°C, as pyrolysis temperature increased to 800°C, the yield of DHBF decreased rapidly to 19.15%. This confirms that in lignin pyrolysis process, the maximum formation of *DHBF* occurred at around 300°C (Lou et al., 2010b). The yield of *E*-isoeugenol and sinapylaldehyde increased slowly because their chemical structures possessed the double bonds of side chain and the benzene rings formed conjugated system, thus became more stable even at high temperature.

4.3 Effect of catalysts on EMAL pyrolysis

The effects of catalysts on the yield of products from EMAL pyrolysis were studied in detail. The quantitative analysis of pyrolysis product based on the integrated areas is shown in Table 3, and the yield distributions of valuable product are present in Fig. 7.

Compound class	Compounds	Yield, Area percent (%)						
		EMAL	EMAL +xNaCl			EMAL +xPermutite		
			5%	10%	20%	5%	10%	20%
Gas	Carbon dioxide	-	-	5.18	2.96	15.21	15.83	18.31
Benzenes	Benzene	-	1.16	1.35	1.59	1.25	1.39	1.82
	Toluene	-	1.83	1.40	2.62	2.13	2.24	2.02
	m-Xylene	-	0.72	0.74	1.25	0.83	1.27	0.94
	Styrene	-	0.95	-	-	0.81	-	-
	Phenol	2.99	3.91	4.40	6.17	5.29	5.64	5.79
p–Hydroxyphenols	o-Cresol	1.68	1.59	2.04	3.20	1.83	2.27	1.89
	p- Cresol	3.58	3.20	3.96	6.03	4.31	5.26	4.65
	o-Allylphenol	-	0.90	2.05	1.11	1.52	2.27	1.35
	m-Xylenol	0.88	1.10	1.71	2.50	1.60	2.07	2.14
	p-Ethylphenol	1.59	2.53	3.15	4.76	2.17	2.78	2.89
	Butylated hydroxytoluene	-	1.04	1.63	1.44	1.00	1.04	1.09
Guaiacols	o-Guaiacol	1.08	0.85	1.85	1.76	1.77	1.27	1.29
	Methoxyl phenol	1.72	3.31	5.70	6.37	1.51	0.81	1.35
	2-Dihydroxytoluene	-	2.60	3.20	4.66	1.74	2.97	1.98
	p-Vinylguaiacol	-	3.84	4.20	4.04	5.15	2.39	4.84
	Vanillin	2.51	0.86	1.20	1.10	1.54	1.99	1.59
	E(Z)-isoeugenol	2.00	2.23	2.55	3.11	2.37	2.22	1.97
	γ-Hydroxyisoeugenol	-	0.85	0.90	0.92	0.66	1.00	1.23
	Acetoguaiacone	-	-	-	-	0.62	0.65	0.83
	Ferulic acid	-	8.40	4.93	3.65	0.60	0.48	0.37
Syringols	Syringol	4.53	1.32	1.99	1.18	5.25	4.94	4.34
	3,4-Dimethoxyphenol	-	1.07	1.70	1.86	-	-	-
	Methoxyeugenol	2.39	1.35	1.02	0.81	2.29	1.66	1.63
	Syringaldehyde	1.45	9.32	2.44	1.25	0.36	0.44	1.51
	Acetosyringone	1.23	-	-	-	0.82	0.86	1.42
	Guaiacylacetone	-	1.56	-	-	-	-	-
Catechols	4-Ethylcatechol	-	-	1.07	1.04	1.11	0.84	0.74
	3-Mmethyl-1,2-benzenediol	-	2.28	3.54	3.77	-	-	-
Heterocycles	2,3-Dihydrobezofuran (*DHBF*)	19.15	16.30	15.43	13.08	21.54	23.78	24.19
Others	Acetic acid	0.97	1.89	3.50	5.84	3.57	3.68	4.01
	Furfural	-	0.97	1.07	1.39	1.08	1.03	0.94
	1,2,4-Trimethoxybenzene	-	1.07	2.01	1.32	2.87	2.34	2.26
	3,5-Dimethoxyacetophenone (*DMAP*)	2.20	3.55	2.52	1.96	3.12	2.22	2.15

Compound class	Compounds	Yield, Area percent (%)						
		EMAL	EMAL +xNaCl			EMAL +xPermutite		
			5%	10%	20%	5%	10%	20%
	p-p′-Isopropylidenebisphenol (*IPBP*)	7.89	2.91	1.38	1.00	-	-	-
	Dibutyl phthalate	1.63	0.90	1.41	0.77	-	-	-
	4-Hydroxy-3,5-dimethoxybenzohydrazide	-	3.62	1.14	0.78	-	-	-

Table 3. Products identification from EMAL pyrolysis with catalysts at 800°C

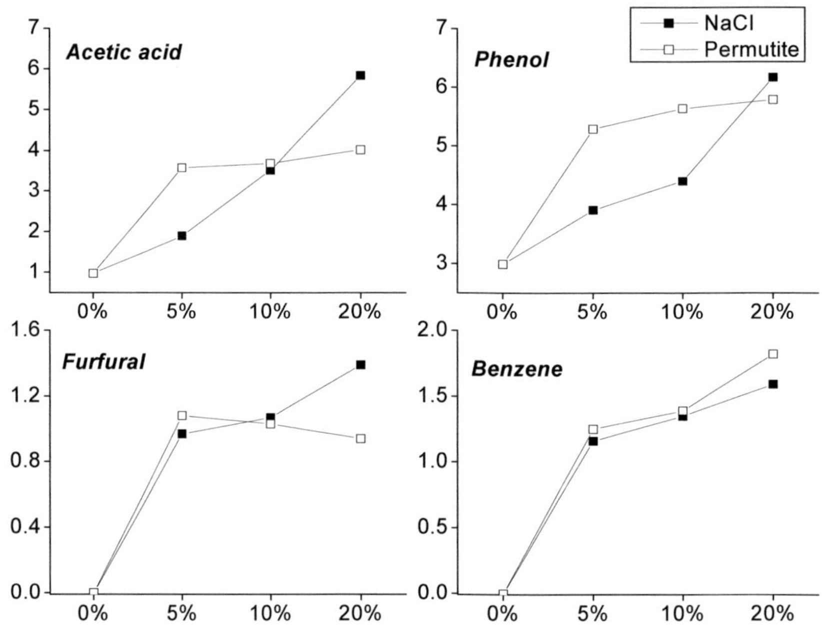

(a)

Fig. 7. Continued

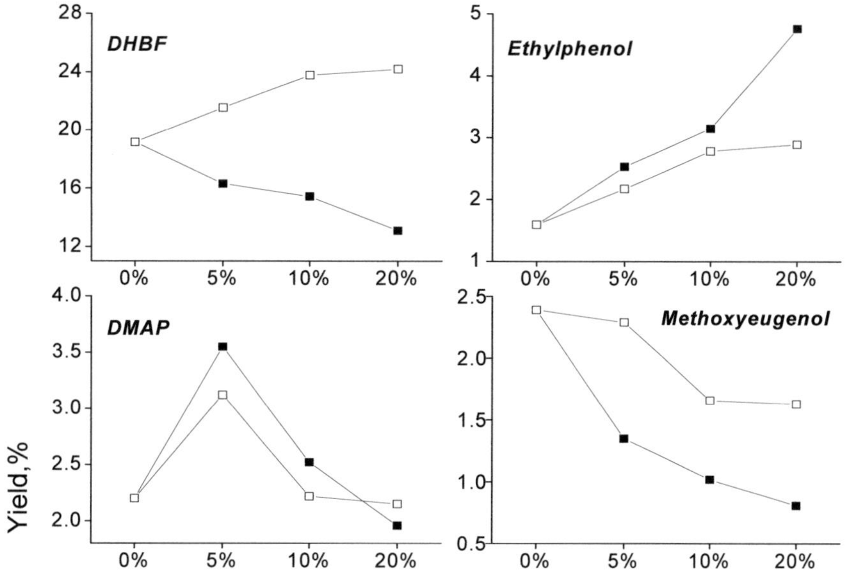

(b)

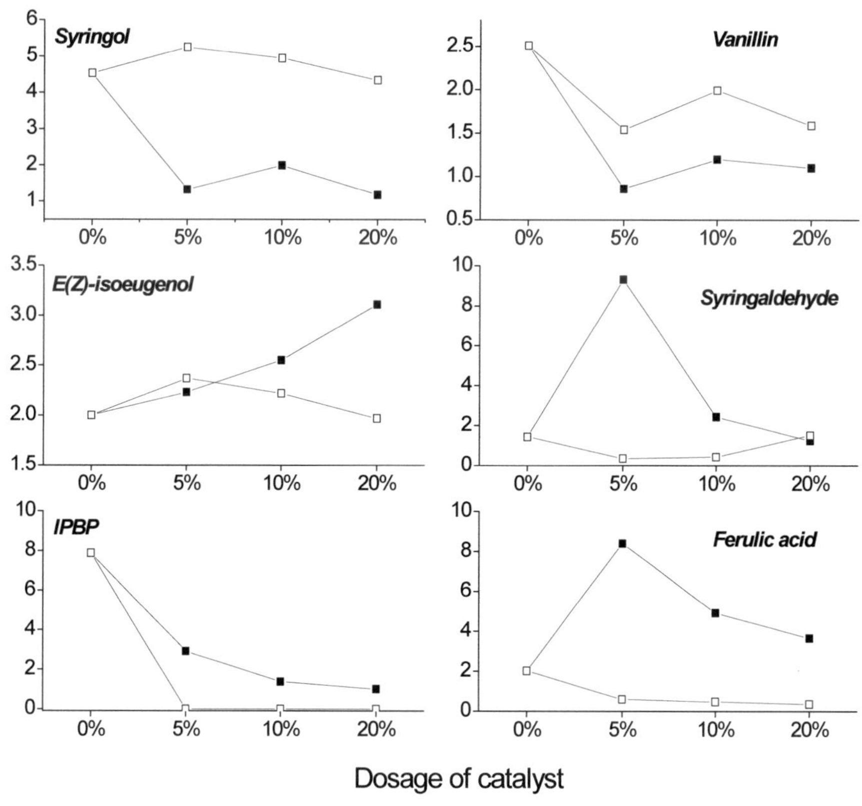

(c)

Fig. 7. Product Yields distribution varied with different dosages of catalyst

Additive Na-salt and permutite promoted the formation of small molecules. The amount of acetic acid, phenol, furfural, and benzene varying with dosage of catalyst is shown in Fig. 7(a). It revealed that yields of acetic acid, phenol, furfural, and benzene increased with an increase of the dosage of two catalysts, and the two catalysts in EMAL pyrolysis promoted the generation of benzene and furfural. When the dosage of permutite was 5%, the formation rate of acetic acid, phenol, and furfural were the most distinct, as the amount of permutite further increased, the increase in yield of them was not obvious. However, as the amount of NaCl increased, the yield enhanced considerably. In short, the addition of two catalysts promoted the cleavage of lignin and the generation of small molecule compounds.

It can be obtained, from Fig. 7(b), that with the additive catalyst increasing from 5% to 20%, permutite has a significant role in promoting the formation of *DHBF* from 19.15% to 24.19%, however, NaCl catalyst was effective to reduce the production of *DHBF* from 19.15% to 13.08%. On the yields of ethylphenol, *DMAP*, and methoxyeugenol during EMAL catalytic pyrolysis, the catalysts of permutite and NaCl had the same impact. Compared with permutite, NaCl catalyst had more pronounced effect to improve or suppress the generation of ethylphenol, *DMAP*, and methoxyeugenol.

Fig. 7(c) shows that when NaCl catalyst was 5%, product of syringaldehyde and ferulic acid reached their maximum yield, while with further increase in the amount of catalyst, the yield decreased. This was similar to that of *DMAP* shown in Fig. 7(b).

Two kinds of phenols' generation trends (i.e. guaiacols and syringols) affected by variations of temperature and catalysts are shown in Fig. 8. Both guaiacols and syringols reached their maximum yields at 600°C. The catalytic effects of NaCl and permutite for improving guaiacols were the most prominent at dosage of 5%, however, the catalytic effect of permutite for syringols was not obvious.

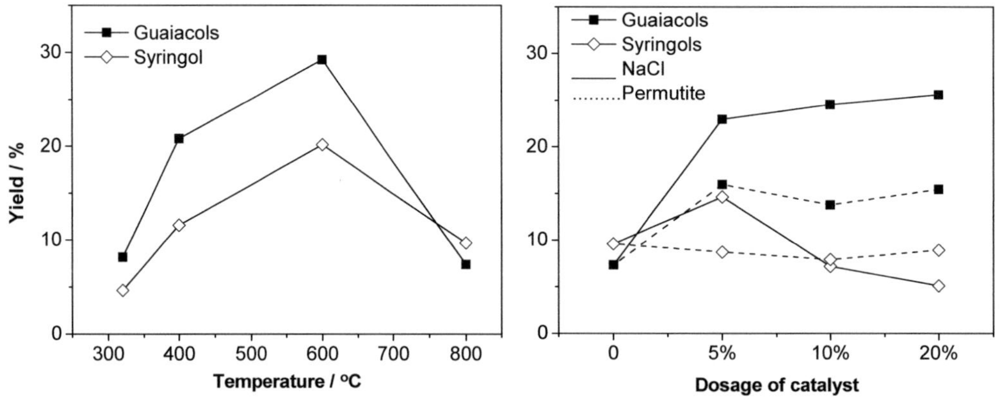

Fig. 8. The total amount of guaiacols and syringols with pyrolysis conditions

To sum up, the impact of different catalysts on the selectivity of pyrolysis products is different. Compared with permutite, the effect of catalyst NaCl was considered to be more significant. Two types of catalysts added to EMAL can promote the generation of small molecule compounds, such as carbon dioxide, acetic acid, benzene series, furfural, and phenol etc. Meanwhile, add catalysts to EMAL made macromolecular lignin degrade to small molecule compounds more thoroughly.

4.4 Effect of temperature on technical lignin pyrolysis

Pyrolysis studies of two types of technical lignin were also started by means of Py-GC/MS analysis. The pyrolysis temperature selection was based on their thermal degradation behaviors (Tan et al., 2009; Wu et al., 2008), respectively.

According to preciously introduced method, identification information of pyrolysis products at different temperatures for two types of technical lignin is presented in Table 4 and 5. Pyrolysis products from technical lignin can be classified into benzenes, phenols, aromatic heterocyclics (mainly 2, 3-Dihydrobenzofuran), esters and trace acids. The lignin derived phenols can be further classified into molecules with guaiacyl, syringyl, and p-hydroxyphenyl aromatic moieties, and were defined as guaiacols, syringols, p-hydroxyphenols, respectively. Yields distribution of each type of product changed with pyrolysis temperature are plotted in Fig. 9.

As the pyrolysis temperature rise, the heterocyclics reduced, while phenolic compounds content increased. The maximum content of 2, 3-Dihydrobenzofuran emerged at about 375°C-400°C. At 600°C, the highest yields of phenols for AL and AHL were 57.91% and 52.11%, respectively. As the temperature increased further, both of their yields were reduced. This characteristic is consistent with that of EMAL.

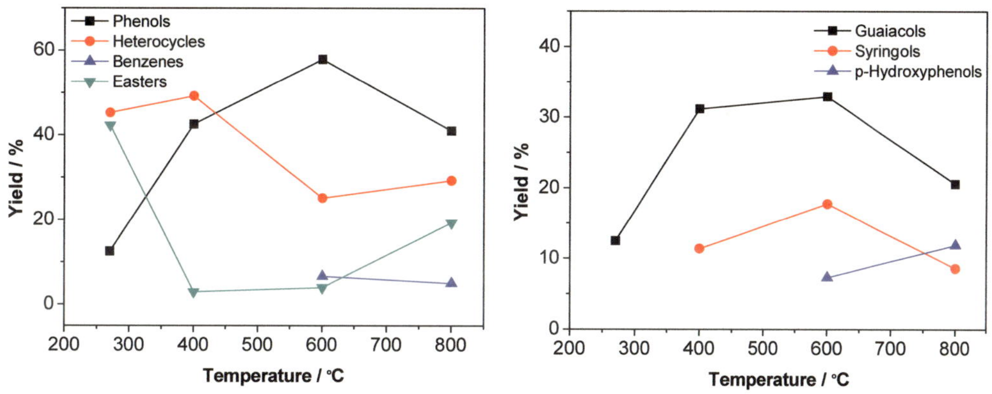

(a) AL

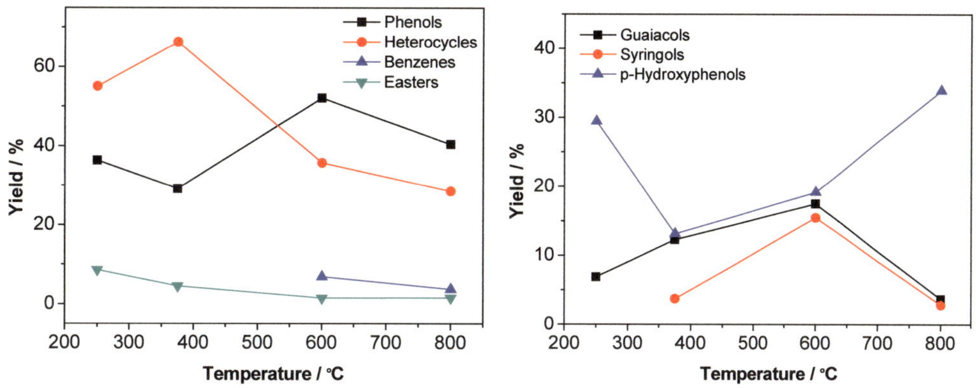

(b) AHL

Fig. 9. Product Yields distribution varied with pyrolysis temperature

Compound class	Compounds	Yield, Area percent (%)			
		270°C	400°C	600°C	800°C
Benzenes	Benzene	-	-	-	0.99
	Toluene	-	-	-	1.06
	1,2,4-Trimethoxybenzene	-	-	6.58	2.89
p–Hydroxyphenols	Phenol	-	-	1.42	3.49
	2-Methylphenol	-	-	1.79	4.49
	4 (or 3)-Ethylphenol	-	-	1.81	2.31
	4-(2-Allyl)-phenol	-	-	2.26	1.58
Guaiacols	2-Methoxyphenol	-	-	6.12	1.58
	2-Methoxy-4-methylphenol	-	-	-	1.44
	3-Methoxycatechol	-	-	3.74	-
	4-Ethyl-2-methoxyphenol	-	-	1.84	0.80
	2-Methoxyl-4-vinylphenol	12.49	27.74	13.60	12.86
	Vanillin	-	-	1.58	-
	2-Mehoxy-3-(2- allyl)-phenol	-	3.43	6.04	1.82
	Coniferylalcohol	-	-	-	2.06
Syringols	2,6-Dimethoxyphenol	-	2.22	7.43	4.87
	Syringaldehyde	-	-	0.95	-
	2,6-Dimethoxy-4-(2-allyl)-phenol	-	9.15	8.00	2.77
	4-Hydroxy-3,5-dimethoxyacetylbenzene	-	-	1.33	0.93
Heterocycles	2,3-Dihydrobenzofuran (*DHBF*)	45.23	49.19	25.09	29.27
Esters	Phenyl glyoxylate-2 'acetyl benzene ester	42.28	2.94	1.76	7.27
	Benzoic acid, phenylmethyl ester	-	-	2.18	10.34
	1,2-Benzenedicarboxylic acid, 1,2-diisooctyl ester	-	-	-	1.72
Others	3,5-Dimethoxy acetophenone	-	5.32	6.77	4.66
	Palmitinicacid	-	-	-	0.80

Table 4. Products identification from AL Pyrolysis

Aromatic esters decreased with temperature increasing, while benzenes compound only emerged at high pyrolysis temperature, with a little increase. Heterocyclic compounds and esters decrease with temperature increasing. This was mainly due to their poor thermal stability, causing these two types of compounds cleaved into smaller molecular, more stable phenols or aromatic compounds at higher temperature.

As about the phenolic products, it can be seen from Fig. 9 that, the guaiacols and syringols trends were similar to that of total phenols, while p–hydroxyphenols had a further increasing trend as the temperature increases. This may be because the branched-chains of the relatively larger molecules of phenols (such as some guaiacols and syringols) cause further cracking as the temperature increases, which resulting the increase of p-hydroxyphenols.

Fig. 10 shows the contents of guaiacols, syringols and p–hydroxyphenols of three types of lignin pyrolyzed at 600°C. As can be seen from the figure, compared with that of AHL, the guaiacols produced from EMAL and AL pyrolysis was much higher (29.21%, 32.92%), while the AHL produced the highest p–hydroxyphenols (19.16%). According to the structural analysis of the three types lignin (Tan, 2009), this product distribution depends qualitatively on the relative content of three basic structure units of lignin.

Compound class	Compounds	Yield, Area percent (%)			
		250°C	375°C	600°C	800°C
Benzenes	Benzene	-	-	-	1.69
	Toluene	-	-	-	2.08
	1,2,4-Trimethoxybenzene	-	-	5.63	-
	1,2,3-Trimethoxyl-5-toluene	-	-	1.30	-
p–Hydroxyphenols	Phenol	-	-	2.83	5.08
	p-Cresol	-	-	3.49	9.35
	4-Methoxylphenol	-	-	3.37	-
	2, 4-Dimethylphenol	-	-	-	1.37
	4 (or 3)-Ethylphenol	-	-	5.18	6.56
	2-Allylphenol	-	-	1.23	2.17
	4-(2-Propenyl)-phenol	-	-	-	4.29
	2,6-Ditert-butyl-4-methyl phenol	29.44	13.15	3.06	5.07
Guaiacols	2-Methoxy-4-methylphenol	-	-	3.64	-
	4-Ethyl-2-methoxyphenol	-	-	2.13	-
	2-Methoxyl-4-vinylphenol	6.91	9.19	7.48	3.71
	Vanillin	-	2.14	1.76	-
	2-Methoxyl-4-(1-propenyl)-phenol	-	0.95	2.48	-
Syringols	2,6-Dimethoxyphenol	-	-	8.19	-
	Syringaldehyde	-	1.23	-	-
	2,6-Dimethoxy-4-(2-allyl)-phenol	-	2.51	7.27	1.16
	4-Hydroxy-3,5-dimethoxyacetylbenzene	-	-	-	1.67
Heterocycles	2,3-Dihydrobenzofuran (DHBF)	55.04	66.28	34.70	26.84
	2, 3-Dihydro-2-methylbenzofuran	-	-	1.00	1.67
Esters	3-Butenevalerate	-	-	-	1.60
	Dibutyl phthalate (DBP)	8.61	4.56	1.50	-
Others	1,3-Butadiene	-	-	-	7.68
	Acetic acid	-	-	-	1.11
	Cyclopentanone	-	-	-	15.68
	3,5-Dimethoxy acetophenone	-	-	3.76	1.23

Table 5. Products identification from AHL Pyrolysis

Lignin pyrolysis producing high value-added phenolic chemicals is a new direction in the future research. The highest content of phenolic compounds occurred at 600°C, meanwhile the content of heterocyclics and esters were minimal, thus had the least effect on the desired phenols. If the temperature was controlled in a reasonable range in the industrial production process, fast pyrolysis of technical lignin could produce more phenols.

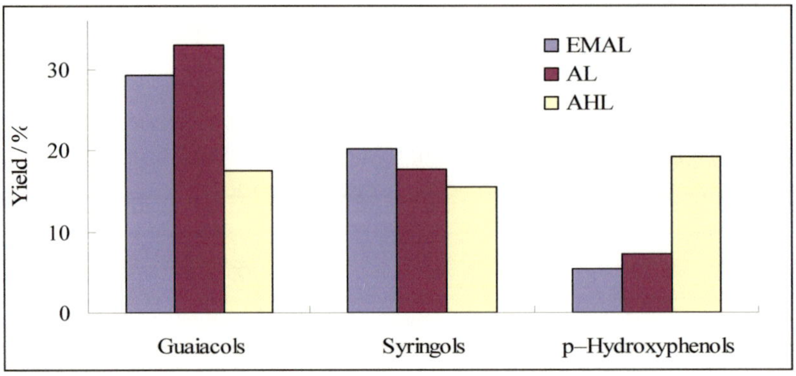

Fig. 10. Phenols contents of three types of lignin pyrolysis at 600°C

4.5 Effect of catalysts on technical lignin pyrolysis

The effects of metal ion catalysts, such as CaO, K_2CO_3, Na_2CO_3, NaCl and $ZnCl_2$, on the pyrolysis reactions of the two types of technical lignin were previously studied by using thermogravimetric analysis and their apparent activation energy were calculated by Coats-Redfern method (Coats & Redfern, 1964). It was found that the effects of these catalysts on both of two types of technical lignin are similar. The apparent activation energy could be reduced by the adding catalysts. In addition, the peak temperatures of lignin decomposition shifted to lower temperature region when adding catalysts, and thus accelerate the lignin pyrolysis reaction. The pyrolysis temperature of two types of technical lignin could be effectively reduced by the addition of K_2CO_3, while the yield of coke pyrolyzed from AL and AHL could be greatly reduced by the addition of NaCl and CaO, respectively.

In order to further study the effect of catalyst on the distribution of pyrolysis products, for each kind of technical lignin, two catalysts, whose effect had been proven to be more obvious by TG analysis, were selected for lignin Py-GC/MS test, i.e. K_2CO_3 and NaCl were selected for AL, while K_2CO_3 and CaO were selected for AHL. In experiment, K_2CO_3 and CaO were mixed with lignin as previously described, while NaCl was used by immersion and adsorption (Tan, 2009).

Products analysis of AL catalytic pyrolysis at 600°C is listed in Table 6, and the product distribution is shown in Fig. 11. It can be seen, the AL pyrolysis products changed dramatically after adding the K_2CO_3 catalyst. The 2, 3-Dihydrobenzofuran (*DHBF*) that possessed the highest content previously was not found in this condition, and the phenols content was reduced to 10.17%. Instead, there have emerged a large number of polycyclic aromatic compounds (are usually exceed two benzene rings of PAHs) and high molecular weight long-chain alkanes compounds. This indicates that K_2CO_3 can effectively promote the cleavage of heterocyclic compounds, but also facilitate further condensation of pyrolysis products, which generate large amounts of naphthalene-based compounds.

The effects of NaCl on lower *DHBF* and phenols content, increase polycyclic aromatic compounds yields are similar to that of K_2CO_3, but its effect is less dramatic. In addition, both two catalysts have led to the rupture of lignin alkyl side-chain, and produced some long-chain alkane compounds.

Compound class	Compounds	Yield, Area percent (%)		
		AL	+10%K_2CO_3	+1%NaCl
Benzenes	1,2,4-Trimethoxybenzene	6.58	3.70	3.06
p-Hydroxyphenols	Phenol	1.42	-	-
	2-Methylphenol	1.79	-	-
	4 (or 3)-Ethylphenol	1.81	-	-
	4-(2-Allyl)-phenol	2.26	-	-
	4-Hydroxy-3-methoxyphenylacetyl	-	3.59	16.14
Guaiacols	2-Methoxyphenol	6.12	3.12	4.88
	2-Methoxy-4-methylphenol	-	-	4.27
	3-Methoxycatechol	3.74	-	-
	4-Ethyl-2-methoxyphenol	1.84	-	2.10
	2-Methoxyl-4-vinylphenol	13.60	-	3.92
	Vanillin	1.58	-	-
	2-Mehoxy-3-(2- allyl)-phenol	6.04	-	-
Syringols	2,6-Dimethoxyphenol	7.43	3.46	6.71
	Syringaldehyde	0.95	-	-
	2,6-Dimethoxy-4-(2-allyl)-phenol	8.00	-	2.35
	4-Hydroxy-3,5-dimethoxyacetylbenzene	1.33	-	-
Heterocycles	2,3-Dihydrobenzofuran (*DHBF*)	25.09	-	21.55
PAH	Phenanthrene	-	7.40	-
	2-Naphthalene acid methyl ester	-	11.98	7.74
	2-Naphthalene-1-butanone	-	7.11	2.98
	1-Phenyl naphthalene	-	11.73	2.88
	2-Acetonaphthone	-	3.99	2.60
	4,5,9,10-Tetrahydropyrene	-	6.17	-
	1,4-2-(2-Dinaphthyl)-butanone	-	7.04	3.39
	4-(2-Naphthyl)-4-ketobutyric acid	-	6.79	-
Esters	Phenyl glyoxylate-2 'acetyl benzene ester	1.76	-	-
	Benzoic acid, phenylmethyl ester	2.18	1.44	-
	4-Methyl-benzoic acid cyclobutyl eater	-	2.99	-
	2,3-dimethyl-2-hexadecylenic acid methyl ester	-	-	1.09
	9-Octadecenoic acid methyl ester	-	-	1.59
Alkanes	n-Nicosane	-	1.61	-
	Hexatriacontane	-	2.42	-
	Triacontane	-	3.90	3.80
Others	Methanol	-	2.08	-
	Docosa-13-en-1-ol	-	2.14	2.48
	1-Triacontanol	-	7.34	6.48
	3,5-Dimethoxy acetophenone	6.77	-	-

Table 6. Products identification from AL pyrolysis with catalysts at 600°C

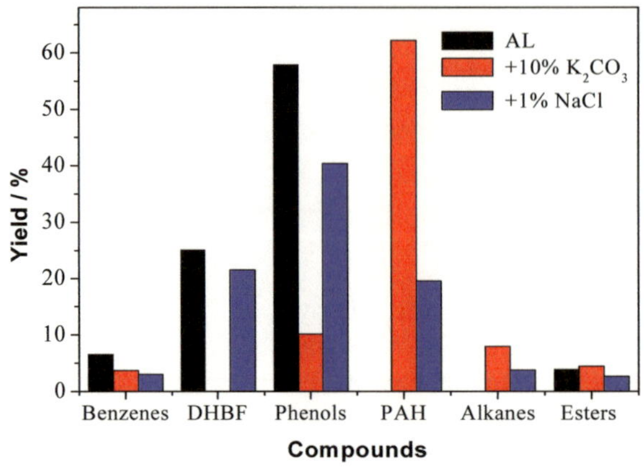

Fig. 11. Effect of catalysts on AL pyrolysis

Table 7 lists the products of AHL pyrolysis with or without catalyst. And Fig. 12 shows the product distribution plotted accordingly. From the figure it can be clearly seen that, the catalytic performance of CaO and K_2CO_3 to AHL pyrolysis was mainly lies in emerging lots of small molecule compounds, particularly for CaO. This is because the catalyst CaO can effectively improve the volatile yield of AHL, and reduce the coke production. In addition, adding CaO also increased the production of heterocyclic compounds, while reducing the phenols content.

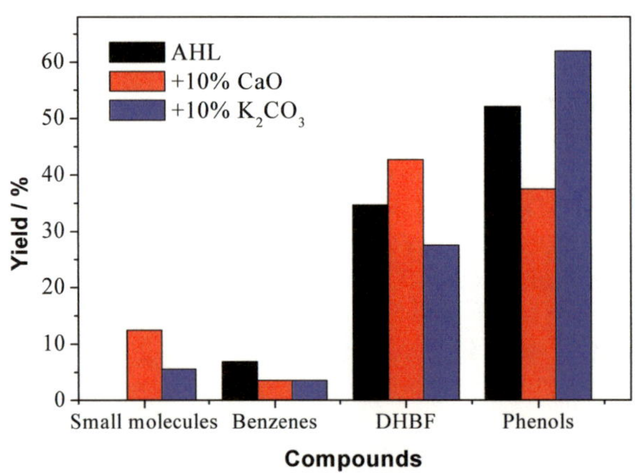

Fig. 12. Effect of catalysts on AHL pyrolysis

In this regard, the effect of K_2CO_3 on AHL pyrolysis is just the opposite. Same as that to AL, K_2CO_3 reduced heterocyclics content of AHL pyrolysis products. The difference is that, for AHL, it did not make further condensation of pyrolysis products to form PAHs, but facilitate the fracture of aromatic compounds side chains, leading to produced more phenolic compounds.

From this it can be seen, as the different structure of lignin, the catalytic effect of a catalyst on different lignins also may be different. The effect of K_2CO_3 on phenolic compounds yield from two technical lignins pyrolysis is a good proof.

Compound class	Compounds	Yield, Area percent (%)		
		AHL	+10%CaO	+10%K_2CO_3
Small molecules	Methanol	-	6.66	-
	Vinyl methyl ether	-	-	1.31
	Methyl acetate	-	-	2.13
	2-Butanone	-	2.78	-
	Glycollic aldehyde	-	1.15	-
	Acetic acid	-	1.91	2.13
Benzenes	Metoxybenzene	-	-	0.74
	1-Vinyl-4-methoxybenzene	-	0.95	-
	1-Ethyl-4- methoxybenzene	-	-	0.49
	1,2-Dimethoxylbenzene	-	-	0.56
	1,2-Dimethoxyl-3-toluene	-	-	1.19
	1,2,4-Trimethoxylbenzene	5.63	2.63	0.63
	1,2,3-Trimethoxyl-5-toluene	1.30	-	-
p-Hydroxyphenols	Phenol	2.83	2.87	9.73
	2-Methyl phenol	-	0.75	1.71
	p-Cresol	3.49	1.86	1.53
	4-Methoxylphenol	3.37	-	-
	4-Vinylphenol	-	-	1.33
	4 (or 3)-Ethylphenol	5.18	2.89	3.44
	2-Allylphenol	1.23	-	-
	4-Hydroxy-3-methylphenylacetyl	-	9.65	9.40
	2,6-Ditert-butyl-4-methyl phenol	3.06	-	-
Guaiacols	2-Methoxyphenol	-	4.06	10.52
	2-Methoxy-4-methylphenol	3.64	4.42	3.20
	4-Ethyl-2-methoxyphenol	2.13	1.66	2.32
	2-Methoxyl-4-vinylphenol	7.48	3.29	0.78
	Vanillin	1.76	-	-
	2-Methoxyl-4-(1-propenyl)-phenol	2.48	-	1.71
Syringols	2,6-Dimethoxyphenol	8.19	4.39	16.31
	2,6-Dimethoxy-4-(2-allyl)-phenol	7.27	1.64	-
Heterocycles	2,3-Dihydrobenzofuran (DHBF)	34.70	42.71	27.52
	2, 3-Dihydro-2-methylbenzofuran	1.00	-	-
Esters	Dibutyl phthalate (DBP)	1.50	-	-
Others	1,4-Dioxane	-	2.52	-
	p-Isopropyl benzaldehyde	-	1.20	1.33
	3,5-Dimethoxy acetophenone	3.76	-	-

Table 7. Products identification from AHL pyrolysis with catalysts at 600°C

5. Conclusion

Pyrolysis of several types of lignins was investigated by using Py-GC/MS, with the focus mainly on the effects of temperatures and catalysts. Significant differences in terms of yields of pyrolysis products and phenolic compounds were observed.

For lignin pyrolysis, temperature is an important parameter, and has a significant impact on both the type and the content of pyrolysis products. High temperature (above 600°C) undoubtedly favored the thermal chemical conversion of lignin. And as the temperature increases, pyrolysis products tend to be more diversified. The 2, 3-Dihydrobenzofuran usually emerged at low pyrolysis temperature and reached its highest yield at about 320°C-400°C. As the temperature increased further, the yield of 2, 3-Dihydrobenzofuran decreased, while the phenolic compounds increased dramatically. At 600°C, the maximum content of phenolic compounds from lignin pyrolysis can reach up to 58%.

The impact of catalysts and its dosage on EMAL pyrolysis mainly reflected in the product distribution, while for technical lignin the impact of catalysts are more evident in the increase of new pyrolysis products. Both NaCl and permutite can promote the generation of small molecule compounds during EMAL pyrolysis, such as acetic acid, benzene series, furfural, and phenols etc., and the yields of those products increase with an increase of catalyst dosage. However, the impact of the two catalysts on the selectivity of 2, 3-Dihydrobenzofuran is opposite. The addition of K_2CO_3 resulted in the dramatically decrease of heterocycles and yielded large amounts of naphthalene-based compounds for the alkali lignin. While the same effect was not happened for the acid hydrolysis lignin, but relatively increase the phenols and small molecules yields.

6. Acknowledgment

This work is supported by the Major State Basic Research Development Program of China (973 Program) (No. 2007CB210201), the National High Technology Research and Development Program of China (863 Program) (No. 2007AA05Z456), and the Natural Science Foundation of China (NSFC, NO. 21176095).

7. References

Adams, R.P. (2007). *Identification of Essential Oil Components By Gas Chromatography/Mass Spectrometry* (4th edition), Allured Pub Corp., ISBN 1-932633-21-9, Illinois, USA

Atika, O., Erika, M., & Rogério, S. et al. (2007). Pyrolysis-GC/MS and TG/MS Study of Mediated Laccase Biodelignification of Eucalyptus Globulus Kraft Pulp. *J. Anal. Appl. Pyrolysis*, Vol.78, No.2, (March 2007), pp. 233-242, ISSN 0165-2370

Baumlin, S., Broust, F., & Bazerbachi, F. et al. (2006). Production of Hydrogen by Lignins Fast Pyrolysis. *International Journal of Hydrogen Energy*, Vol.31, No.15, (December 2006), pp. 2179-2192, ISSN 0360-3199

Bridgwater, A.V., Meier, D. & Radlein, D. (1999). An Overview of Fast Pyrolysis of Biomass. *Organic Geochemistry*, Vol.30, No.12, (December 1999), pp. 1479-1493, ISSN 0146-6380

Bridgwater, A.V. (2004). Biomass Fast Pyrolysis. *Thermal Science*, Vol.8, No.2, (March 2004), pp. 21-50, ISSN 0354-9836

Britt, P.F., Buchanan, A.C. & Thomas K.B. et al. (1995). Pyrolysis Mechanisms of Lignin: Surface-Immobilized Model Compound Investigation of Acid-Catalyzed and Free-Radical Reaction Pathways. *J. Anal. Appl. Pyrolysis*, Vol.33, (April 1995), pp. 1-19, ISSN 0165-2370

Bull I.D. (April 2005). Pyrolysis gas chromatography mass spectrometry (Py/GC/MS), In: *Techniques*, 27.04.2005, Available from http://www.bris.ac.uk/nerclsmsf/techniques/pyro.html

Coats, A.W. & Redfern, J.P. (1964). Kinetic Parameters from Thermogravimetric Data. *Nature*, Vol.201, (January 1964), pp. 68-69, ISSN 0028-0836

Domínguez, J.C., Oliet, M., & Alonso, M.V. et al., (2008). Thermal Stability and Pyrolysis Kinetics of Organosolv Lignins Obtained from Eucalyptus Globulus. *Industrial Crops and Products*, Vol.27, No.2, (March 2008), pp. 150-156, ISSN 0926-6690

Fahmi, R., Bridgewater, A.V. & Thain, S.C. (2007). Prediction of Klason Lignin and Lignin Thermal Degradation Products by Py–GC/MS in a Collection of Lolium and Festuca Grasses. *J. Anal. Appl. Pyrolysis*, Vol.80, No.1, (August 2007), pp. 16-23, ISSN 0165-2370

Fu, R.N. (2008). *Overview of Chromatographic Analysis* (2nd edition), Chemical Industry Press, ISBN 9787502564988, Beijing, China

Garcia, P.M., Wang, S. & Shen, J. et al. (2008). Effects of Temperature on The Formation of Lignin-derived Oligomers during the Fast Pyrolysis of Mallee Woody Biomass. *Energy & Fuels*. Vol.22, No.3, (March 2008), pp. 2022–2032, ISSN 0887-0624

Harumi, H., Satoshi, K. & Tatsuhiko, Y. et al. (2010). Conversion of Technical Lignins to Amphiphilic Derivatives with High Surface Activity. *Journal of Wood Chemistry and Technology*, Vol.30, No.2, (May 2010), pp. 164-174, ISSN 0277-3813

Hosoya, T., Kawamoto, H. & Saka, S. (2007). Pyrolysis Behaviors of Wood and Its Constituent Polymers at Gasification Temperature. *J. Anal. Appl. Pyrolysis*, Vol.78, No.2, (March 2007), pp. 328-336, ISSN 0165-2370

Jegers, H.E. & Klein, M.T. (1985). Primary and Secondary Lignin Pyrolysis Reaction Pathways. *Ind. Eng. Chem. Process Des. Dev.*, Vol.24, No.1, (January 1985), pp. 173-183, ISSN 0196-4305

Jin, X.G., Huang, L.Y. & Shi, Y. (2009). *Pyrolysis Gas Chromatography-Method and Application* (1st edition), Chemical Industry Press, ISBN 9787122046567, Beijing, China

Lee, G.R. & Eugene F. B. (2004). *Modern Practice of Gas Chromatography* (4th edition), Wiley-Interscience, ISBN 0-471-22983-0, New Jersey, USA

Liu, Q., Wang, S.R. & Zheng, Y. et al., (2008). Mechanism Study of Wood Lignin Pyrolysis by using TG - FTIR Analysis. *J. Anal. Appl. Pyrolysis*, Vol.82, No.1, (May 2008), pp. 170-177, ISSN 0165-2370

Lou, R. & Wu, S.B. (2008). Pyrolysis Characteristics of Rice Straw EMAL. *Cellulose Chem. Technol.*, Vol.42, No.7-8, (December 2008), pp. 371-380, ISSN 0576-9787

Lou, R., Wu, S.B. & Lv, G.J. (2010a). Fast Pyrolysis of Enzymatic/Mild Acidolysis Lignin from Moso Bamboo. *BioResources*, Vol.5, No.2 (March 2010), pp. 827-837, ISSN 1930-2126

Lou, R., Wu, S.B. & Lv, G.J. (2010b). Effect of Conditions on Fast Pyrolysis of Bamboo Lignin. *J. Anal. Appl. Pyrolysis*, Vol.89, No.2, (November 2010), pp. 191-196, ISSN 0165-2370

Lou, R. & Wu, S.B. (2011). Products Properties from Fast Pyrolysis of Enzymatic/Mild Acidolysis Lignin. *Applied Energy*, Vol.88, NO.1, (January 2011), pp. 316-322, ISSN 0306-2619

Lv, G.J., Wu, S.B., & Lou, R. (2010a). Kinetic Study of the Thermal Decomposition of Hemicellulose Isolated from Corn Stalk. *Bioresources*, Vol.5, NO.2, (April 2010), pp. 1281-1291, ISSN 1930-2126

Lv, G.J., Wu, S.B., Lou, R. & Yang, Q. (2010b). Analytical Pyrolysis Characteristics of Enzymatic/Mild Acidolysis Lignin from Sugarcane Bagasse. *Cellulose Chem. Technol.*, Vol.44, NO.9, (October 2010), pp. 335-342, ISSN 0576-9787

Meier, D. & Faix, O. (1999). State of the Art of Applied Fast Pyrolysis of Lignocellulosic Materials-a Review. *Bioresource Technology*, Vol.68, No.1, (April 1999), pp. 71-77, ISSN 09608524

Mistrik, R. (2004). A New Concept for the Interpretation of Mass Spectra Based on a Combination of a Fragmentation Mechanism Database and a Computer Expert System, In: *Advances in Mass Spectrometry*, Brenton, G., Monaghan, J. & Ashcroft, A., pp. 821-824, Elsevier, ISBN 0-444-51528-3, Amsterdam

Mohan, D., Pittman, C.U. & Steele, P.H. (2006). Pyrolysis of Wood/Biomass for Bio-oil: A Critical Review. *Energy & Fuels*, Vol.20, No.3, (March 2006), pp. 848–889, ISSN 0887-0624

Nowakowski, D.J. & Jones, M.J. (2008). Uncatalysed and Potassium-Catalysed Pyrolysis of the Cell-Wall Constituents of Biomass and Their Model Compounds. *J. Anal. Appl. Pyrolysis*, Vol.83, No.1, (September 2008), pp. 12-25, ISSN 0165-2370

Nowakowski, D.J., Bridgwater, A.V. & Elliott, D.C. et al. (2010). Lignin Fast Pyrolysis: Results from an International Collaboration. *J. Anal. Appl. Pyrolysis*, Vol.88, No.1, (May 2010), pp. 53–72, ISSN 0165-2370

Kealey, D. & Haines P. J. (2002). *Analytical Chemistry*, BIOS Scientific Publishers Ltd, ISBN 1-85996-189-4, Oxford, UK

Sparkman, O.D. (2000). *Mass Spectrometry Desk Reference*, Global View Pub, ISBN 0-9660813-2-3, Pittsburgh, USA

Tan, Y. (2009). *Study on the Chemical Structures and Thermochemical Laws of Two Types of Industrial Lignin*, South China University of Technology, Doctoral Dissertation, Guangzhou, China

Tan, Y., Wu, S.B., & Lou, R. et al. (2009). Thermogravimetric Characteristics and Kinetic Analysis of Lignin Hydrolyzed by Dilute Acid. *Journal of South China University of Technology (Natural Science)*, Vol.37, No.6, (June 2009), pp. 22-26, ISSN 1000-565X

Wang, S.R., Wang, K.G. & Liu Q. et al. (2009). Comparison of the Pyrolysis Behavior of Lignins from Different Tree Species. *Biotechnology Advances*, Vol.27, No.5, (September 2009), pp. 562-567, ISSN 0734-9750

Windt, M., Dietrich, M., & Jan, H.M. et al. (2008). Micro-pyrolysis of Technical Lignins in a New Modular Rig and Product Analysis by GC-MS/FID and GC×GC-TOFMS/FID. *J. Anal. Appl. Pyrolysis*, Vol.85, No.1-2, (May 2009), pp. 38-46, ISSN 0165-2370

Wu, L.J. (2005). *Gas Chromatography Detection Method* (1st edition), Chemical Industry Press, ISBN 978750256953, Beijing, China

Wu, S.B. & Argyropoulos, D.S. (2003). An Improved Method for Isolating Lignin in High Yield and Purity. *Journal of Pulp and Paper Science*, Vol.29, NO.7, (July 2003), pp. 235-240, ISSN 0826-6220

Wu, S.B., Xiang, B.L., & Liu, J.Y. et al. (2008). Pyrolysis Characteristics of Technical Alkali Lignin. *Journal of Beijing Forestry University*, Vol.30, NO.5, (September 2008), pp. 143-147, ISSN 1000-1522

Yang, Q., Wu, S.B., Lou, R. & Lv, G.J. (2010). Analysis of Wheat Straw Lignin by Thermogravimetry and Pyrolysis-Gas Chromatography/Mass Spectrometry. *J. Anal. Appl. Pyrolysis*, Vol.87, No.1, (January 2010), pp. 65-69, ISSN 0165-2370

4

Degradation Phenomena of Reforming Catalyst in DIR-MCFC

Kimihiko Sugiura
Osaka Prefecture University College of Technology, Japan

1. Introduction

1.1 Background

Recently, energy consumption is increasing with the population growth of developed countries and developing countries, and an exhausting of petroleum resources has become a big problem worldwide. In addition, CO_2, SO_X and NO_X which are toxic substances emitted when these resources are used, cause environmental polluting problem such as global warming and acid rain, etc. From these situations, high efficiency, energy-saving and environmental-saving power source is needed in recent years, and fuel cell can solve these problems. Therefore, researching, developing and commercializing on fuel cell are promoted in many countries. Fuel cell is categorized by the difference of its electrolyte and operating temperature. In this research, we use Molten Carbonate Fuel Cells (MCFC) which uses a molten carbonate as the liquid electrolyte. MCFC is expected as a distributed power station and the CO_2 concentrator .Generally, fuel cell performances are evaluated by I-V performance, cell resistance and gas chromatography, etc. for which fuel cell only generate by supplying fuel gas and oxidant gas to cell. Generally, although gas chromatography is utilized only to detect the gas crossover (it is gas leakage phenomenon between two poles through electrolyte) on fuel cell development, the performance of reforming catalyst in the anode channel in operation is evaluated by gas chromatography on R&D of Direct Internal Reforming-Molten Carbonate Fuel Cell (DIR-MCFC). Moreover, MCFC installed the segmented electrode is evaluated about the gas crossover and the utility of the segmented electrode by a gas chromatography and the image measurement technique. This chapter introduces the use method of a gas chromatograph through the degradation factor of reforming catalyst in DIR-MCFC.

1.2 Kind of fuel cell, power generation principle and Issues of DIR-MCFC

From above-mentioned, the kind of fuel cell is decided by the electrolyte used, and the operating temperature, the fuel used and the system structure are decided as shown in Table 1. For example, although Phosphoric Acid Fuel Cell (AFC) and Polymer Electrolyte Fuel Cell (PEFC) exhaust the produced water to the cathode side, MCFC, Solid Oxide Fuel Cell (SOFC) and Alkaline Fuel Cell (AFC) exhaust it to the anode side. Therefore, the system structure is different by fuel cell used. On the fuel gas, the high temperature operation type fuel cell such as MCFC and SOFC can use various fuels, because they do not use an electrode catalyst and they can use CO as a fuel. However, the low temperature operation type fuel cell such as PAFC and PEFC use only hydrogen because CO becomes a catalyst poison, and AFC can only, moreover, pure hydrogen because CO_2 makes the electrolyte be deteriorated. Therefore, the low temperature operation type fuel cell needs an addition facility such as the shift convertor.

	AFC	PAFC	MCFC	SOFC	PEFC
Electrolyte	Potassium hydroxide	Phosphoric acid	Molten carbonate	Stabilized zirconium	Ion-exchange membrane
Ionic conductor	OH^-	H^+	CO_3^{2-}	O^{2-}	H^+
Specific resistance	< 1Ωcm	< 1Ωcm	< 1Ωcm	< 1Ωcm	< 20Ωcm
Operating temperature	50 - 150°C	190 - 220°C	600 - 700°C	- 1000°C	80 - 120°C
Corrosively	Medium	Strong	Strong	non	Medium
Electrode catalyst	Nickel/silver	Platinum	Not need	Not need	Platinum
Fuel	Pure H_2 without CO_2	H_2 (CO_2 content; OK)	H_2 & CO	H_2 & CO	H_2 (CO_2 content; OK)
Fuel source	By-product hydrogen of electrolytic industry, Water electrolysis	Natural gas, Light oil, Methanol	Oil, Natural gas, Methanol, Coal gas	Oil, Natural gas, Methanol, Coal gas	Natural gas, Methanol
Generating efficiency	60%	40 – 50%	40 – 60%	50 – 65%	40 – 50%

Table 1. Kind and feature of fuel cell

Because this study targets MCFC, the power generation principle of MCFC is shown in Fig.1. The fuel of MCFC is hydrogen obtained by reforming the natural gas and the coal gasification gas, and CO_2 that stabilizes the molten salt and steam that prevents the carbon deposition are contained in the fuel gas. Hydrogen supplied to the cell is separated to the proton and the electron with the anode catalyst ($H_2 \rightarrow 2H^+ + 2e^-$). This proton and carbonate ion, which has been transmitted in electrolyte, generate steam and CO_2 with the electrochemical reaction ($2H^+ + CO_3^{2-} \rightarrow H_2O + CO_2$).

On the other hand, oxygen supplied to the cell and electrons transmitted from external circuit react into oxygen ion with the cathode catalyst ($0.5\ O_2 + 2e^- \rightarrow O^{2-}$). Carbon dioxide supplied to the cell deliquesces to the electrolyte on the cathode electrode, and it becomes carbonate ion as this CO_2 reacts with oxygen ion ($CO_2 + O^{2-} \rightarrow CO_3^{2-}$). Here, the carbonate ion generated is the electrolyte, and it moves toward the anode electrode. Overall reaction is an evolution reaction of water.

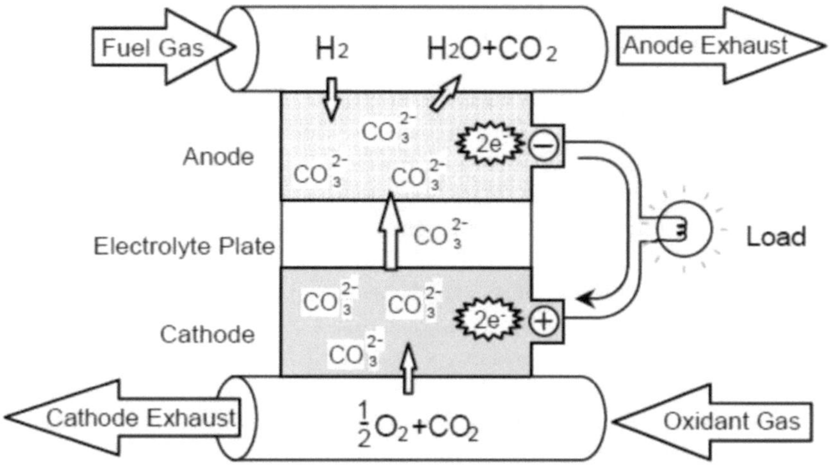

Fig. 1. Schematic diagram of power generation principle of MCFC

MCFC has three reforming type to obtain hydrogen from various fuel as shown in Fig.2. The kinds of reforming type on MCFC are classified into the external reforming type, which has the reformer outside the cell as shown in Fig.2 (a), and the internal reforming type, which

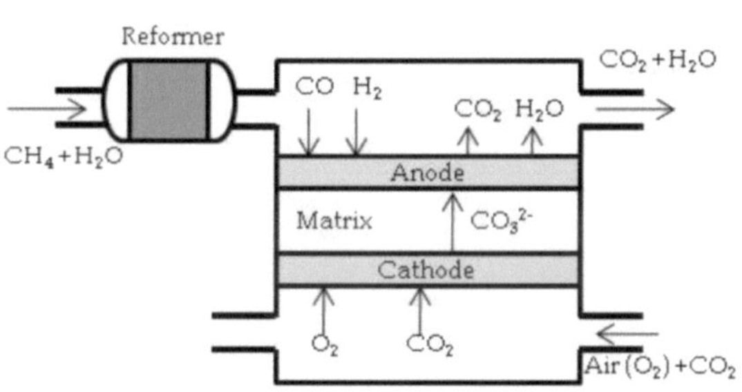

(a) External Reforming type

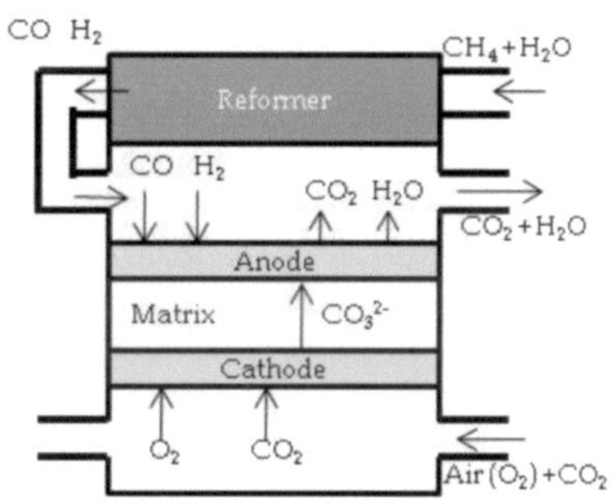

(b) Indirect internal Reforming type

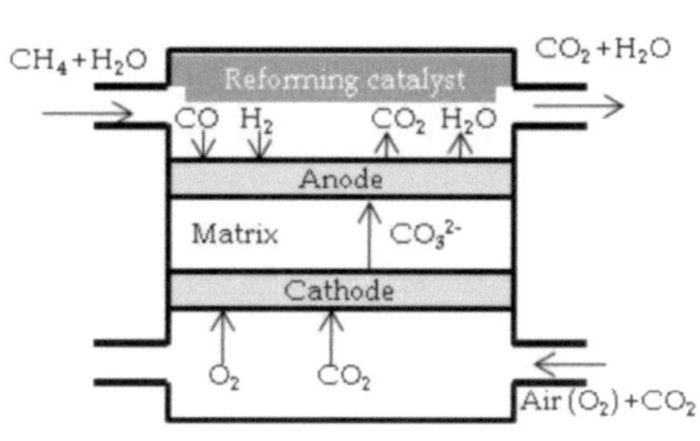

(c) Direct internal Reforming type

Fig. 2. Schematic diagram of comparison of reforming type

has the reformer inside the cell. Moreover, the internal reforming type is classified into the indirect reforming type, which sets up the reformer among cells as shown in Fig.2 (b), and the direct internal reforming type, which fills the reforming catalyst directly to the anode channel as shown in Fig.2 (c). Although the merits of ER-MCFC (Fig.2 (a)) are a simple structure and the enlargement of MCFC is easy, the demerits are that a big space required for installation and heat supply to reformer are necessary. The merits of IIR-MCFC (Fig.2 (b)) can make the system compact by stacking the reformer of one every five cells, and can control the cell temperature by the endothermic reaction of the reformer. However, as a demerit, the system and the cell structure of IIR-MCFC become more complex than that of ER-MCFC, and the height of IIR-MCFC stack becomes taller than ER-MCFC only by the thickness of the reformer. On the other hand, because DIR-MCFC (Fig.2 (c)) has a reforming catalyst in the anode channel, the system size of DIR-MCFC is smaller than that of ER-MCFC, and the cell structure is simple as well as ER-MCFC. Moreover, the reforming reaction is promoted from the chemical equilibrium value because MCFC consumes hydrogen to which the reforming catalyst reforms by the cell reaction. Therefore, the generation efficiency of DIR-MCFC is larger than that of ER-MCFC and IIR-MCFC.

Because MCFC uses the electrolyte of liquid phase, it has the problem that the electrolyte decrease as the operating time passes (Tanimoto, K. et al., 1998). However, because the reforming catalyst is loaded in the anode gas channel, the possibility that it is polluted by the molten salts of liquid phase and vapour phase is high as shown in Fig.3. The liquid phase pollution is caused by which the molten salt leak through the cell component such as separators and adhere to the catalyst (Sugiura et al., 1999). The vapour phase pollution is caused by which the vaporised electrolyte disperse to the gas channel with the steam generated by the cell reaction and adhere to the catalyst. The reforming catalyst is polluted by adhering of these molten salts, and the performance of the reforming catalyst deteriorates

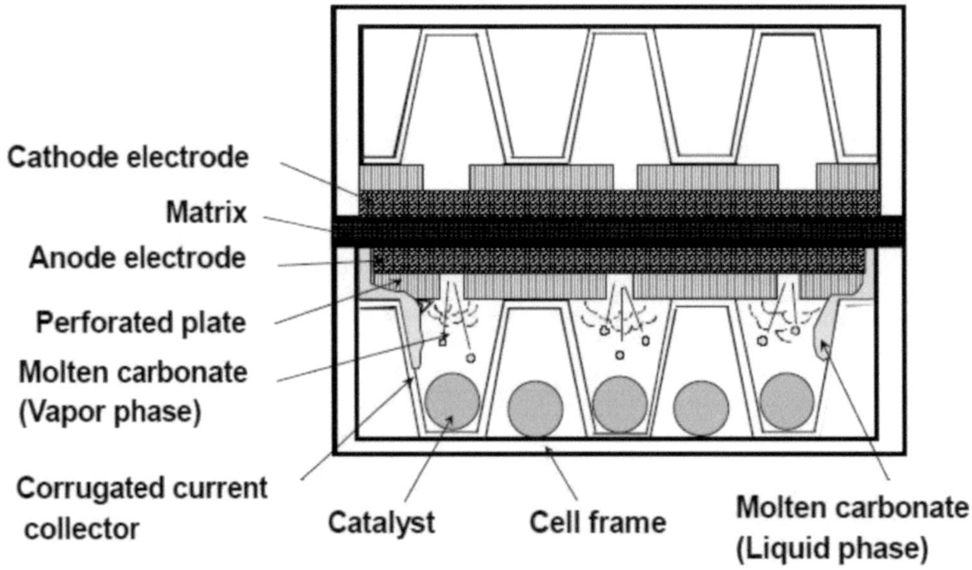

Fig. 3. Schematic diagram of catalyst pollution mechanism by molten salts in DIR-MCFC

and DIR-MCFC cannot generate electricity finally. Therefore, these pollution phenomena of the reforming catalyst should be confirmed by the DIR-MCFC single cell, and the prevention method of the catalyst pollution should be proposed.

2. Experimental

2.1 Experimental apparatus and conditions

Figure 4 shows the schematic diagram of the experimental apparatus. The heater plate was set up in the top and bottom of the DIR-MCFC single cell, and the cell temperature was maintained at 650°C by which the DIR-MCFC single cell be blanketed by the insulation using the thermo-regulator (E5T@Omron, Japan). The reduction of the cell contact resistance and the gas sealing in the cell were carried out by compressing the single cell, the heater plates and the insulation by air press. H_2 and CO_2 as anode gases were adjusted by the mass flow controller (3660@Kofloc, Japan) and mixed with the mixer that is contained glass ball, and were humidified by passing through a humidifier and being supplied to DIR-MCFC. To prevent the dew formation of the supply gas, the piping between the humidifier and the cell was made shorter, and the piping temperature was maintained by a ribbon heater at 120°C. On the other hand, air from the compressor and CO_2 as cathode gas were also adjusted by the mass flow controller and mixed with the mixer, and were supplied to DIR-MCFC. Cell voltage was measured and recorded by a data logger (DR130@Yokogawa Electric Co. Ltd, Japan). Cell resistance was measured by a milliohm meter with AC 4 probes (3566@Turuga Electric Co. Ltd, Japan).

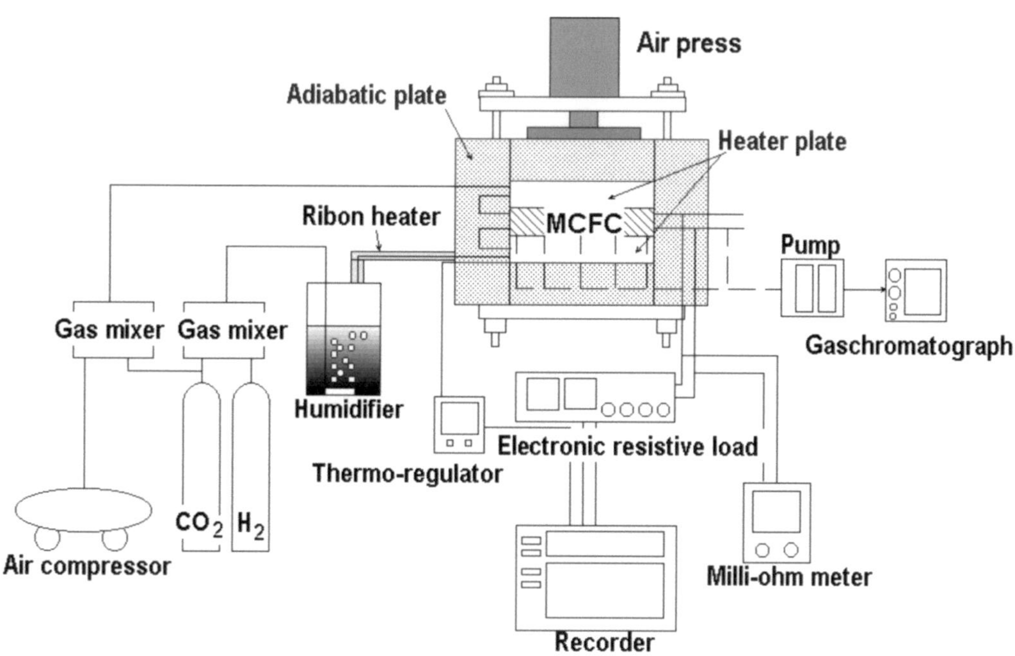

Fig. 4. Schematic diagram of experimental apparatus

The DIR-MCFC single cell had five sampling port for measuring the gas composition distribution in the cell as shown in Fig.5, and the sampling gas was collected by the pump through these ports, and analyzed with the gas chromatograph (GC-8AIT@Shimazu Co., Japan; the detector was the thermal conductivity detector). The locations of each sampling port were 3, 25, 50, 75 and 97% against the direction of flow, respectively. The reforming catalyst was loaded into the valley of the anode corrugated current collector as shown in Fig.6. Here, the corrugated current collector of anode side was made of Inconel steel, and that of cathode side was made of SUS316L, respectively. The standard gas condition was called ER mode; anode gas composition was $72H_2/18CO_2/ 10H_2O$ (Volume %), cathode gas composition was $70Air/30CO_2$, and the gas utilization of both sides was 40%, respectively. The standard condition of DIR-MCFC was called DIR mode; anode gas composition was $25CH_4/75H_2O$ (steam carbon ratio is 3.0), and cathode composition was same as ER mode. The operating atmosphere was 0.1MPa. As shown in Fig 3, general MCFC single cell was composed of anode cell frame (SUS316L), anode corrugated current collector (Inconel), anode current collector (Ni), anode electrode, matrix, cathode electrode (NiO), cathode

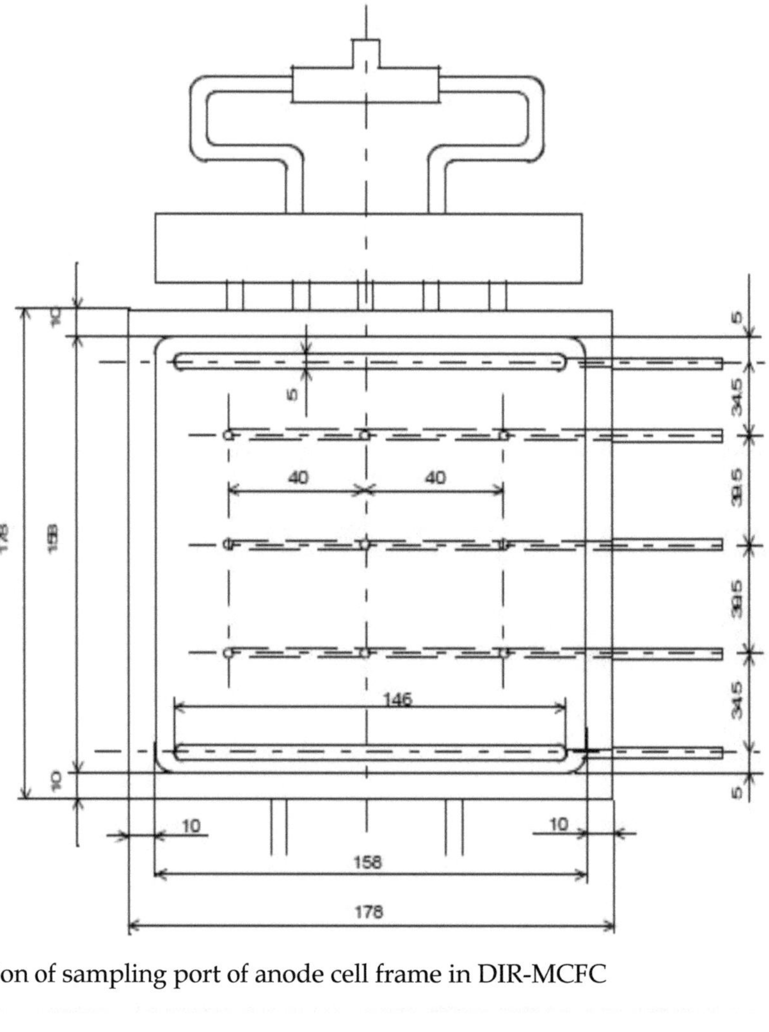

Fig. 5. Location of sampling port of anode cell frame in DIR-MCFC

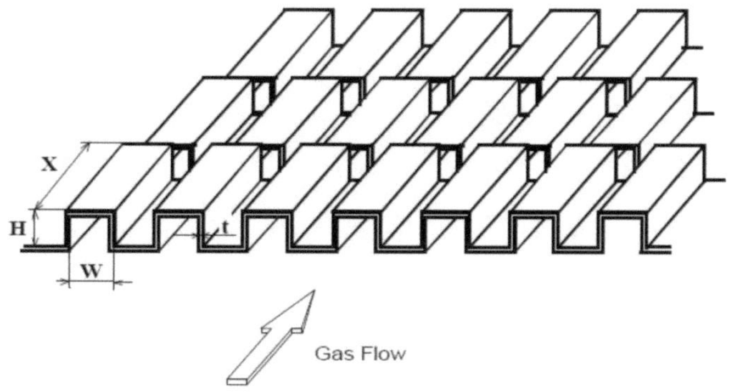

Fig. 6. Structure of corrugated current collector

current collector (SUS316L), cathode corrugated current collector (SUS316L) and cathode cell frame (SUS316L). Table 2 shows the specification of each cell component. All cell components such as anode electrode, cathode electrode, matrix and electrolyte sheet were made by a doctor blade method. The anode electrode was made of Ni-3wt%Al powder. The cathode electrode was made of Ni powder as a green sheet, and this green sheet was changed to NiO in the cell while the start up procedure. The matrix was made of $LiAlO_2$. The electrolyte was made of $62Li_2CO_3/38K_2CO_3$.

	Anode	Cathode	Matrix
Materials	Ni-3wt%Al	NiO (in situ)	$LiAlO_2$
Thickness [mm]	0.75-0.85	0.75-0.85	0.50
Porosity [%]	50	55	50
Mean pore diameter [micron]	3	1-4, 0.2-0.4	0.1

Table 2. Specification of cell components

2.2 Setting of gas chromatograph

In fuel cell research, the crossover between anode and cathode through the electrolyte and the activation of the reforming catalyst are evaluated by the gas analysis. Especially, because CH_4 is reformed to H_2 in the anode channel of DIR-MCFC, H_2, CH_4, CO and CO_2 concentration become the index of the catalytic activity. Moreover, because neither N_2 nor O_2 exist in the anode gas, N_2 and O_2 become the index of the crossover. To analyse these gases in short time, the column diverged to two passages, one passage was composed by 0.5m of the shimalite Q 100/180 and 1.5m of the porapak Q 80/100, and another passage was composed by 2.5m of the molecular sieve 5A 60/80 as shown in Fig.7. The carrier gas was Argon gas, the flow rate of the carrier gas was 50ml/min, and the inlet pressure was 121.6kPa, respectively. The temperature of the column, the detector and the injection temperature were 90°C, 100°C and 100°C, respectively. Figure 8 shows the chromatogram obtained by this column. The peak of CO_2 does not overlap with that of air because air is hardly contained in the anode gas though the peak of air appears immediately after the analysis beginning. Because the desorption of CO is slow, the analysis time is about 20 minutes.

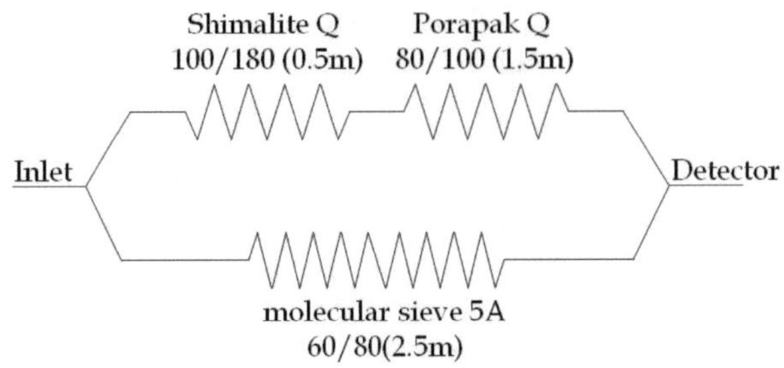

Fig. 7. Schematic diagram of the column structure for DIR-MCFC

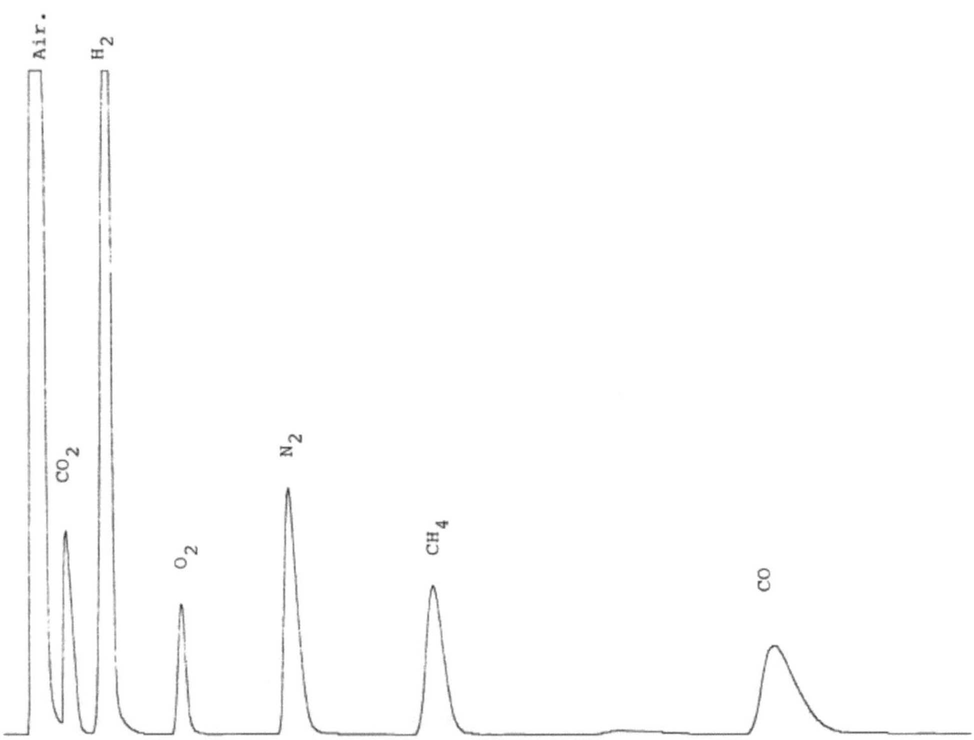

Fig. 8. Example of chromatogram using the column for DIR-MCFC

2.3 Loading method of reforming catalyst in the anode channel

The influence of the loading method of the reforming catalyst on the cell performance is examined by four loading methods as shown in Fig.8. Here, the reforming catalyst was the commercially available nickel catalyst (25wt%Ni/MgO; Type RKNR@ Haldor Topsoe, Denmark), and is 1.2mmO.D.x1.4mmL of the cylinder type.

a. "Frame side" means that the reforming catalyst was loaded into only the frame side of the corrugated current collector to keep away the catalyst from the electrode that contains the carbonate as shown in Fig.9 (a). The catalyst loading density was 48mg/cm^2.

b. "Electrode side" means that the reforming catalyst was loaded into only the electrode side of the corrugated current collector to improve the reforming rate by making the reforming reaction adjoin the cell reaction as shown in Fig.9 (b). The catalyst loading density was 78mg/cm^2.

c. "Upper stream" means that the reforming catalyst was loaded into both side of the upper stream part and the frame side of the downstream part of the corrugated current collector to cool the vicinity of the cathode outlet locally as shown in Fig.9 (c). The catalyst loading density of the frame side and the electrode side were 60mg/cm^2 and 30mg/cm^2, respectively.

d. "Downstream" means that the reforming catalyst was loaded into both side of the downstream part and the frame side of the upper stream part of the corrugated current collector to cause the reforming reaction uniformly with the entire anode electorde as shown in Fig.9 (d). The catalyst loading density of the frame side and the electrode side were 60mg/cm^2 and 30mg/cm^2, respectively.

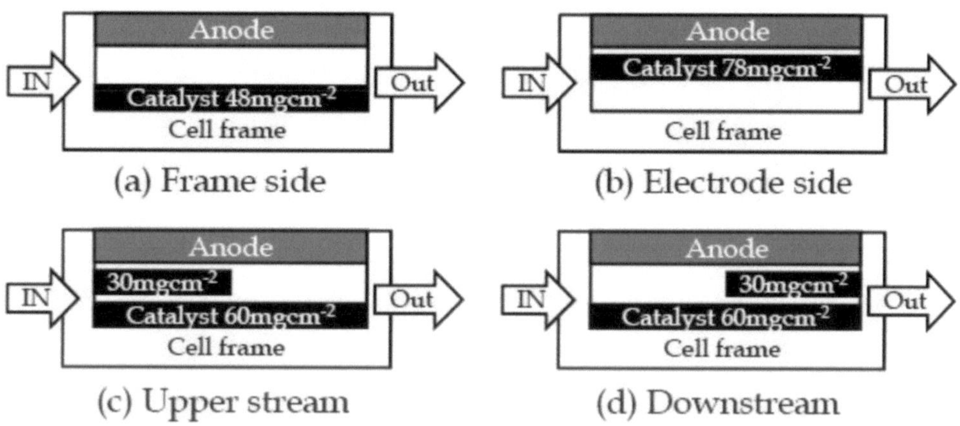

Fig. 9. Schematic diagram of loading method of the reforming catalyst and loading amount

3. Results and discussion

3.1 Cell performance

Figure 10 shows the initial performance comparison between the ER mode (the reforming catalyst does not work) and the DIR mode under the downstream of loading condition. Because the difference of both modes is little, the reforming catalyst works correctly at the initial period. Figure 11 shows the influence of the loading method of the reforming catalyst on the life performance. Here, this result is the cell performance under DIR mode. Although all cell voltages are almost same until 300 hours of the operating time, the cell performances other than the cell of "downstream" are rapidly deteriorated. Moreover, the cell performance of "downstream" is also deteriorated when the operating time exceeds 600 hours. This deteriorating reason should be judged the catalyst pollution originating or the cell deteriorating originating. Therefore, the deteriorating reason is evaluated by comparing of the performance under ER mode with that of the "Downstream" cell under DIR mode as shown in Fig.12. Although the cell performance under ER mode is good, the cell performance under DIR mode deteriorates gradually from 100 hours to 600 hours of operating time, and deteriorates rapidly when the operating time exceeds 600 hours. It is a peculiar problem to DIR-MCFC (Miyake et al., 1995).

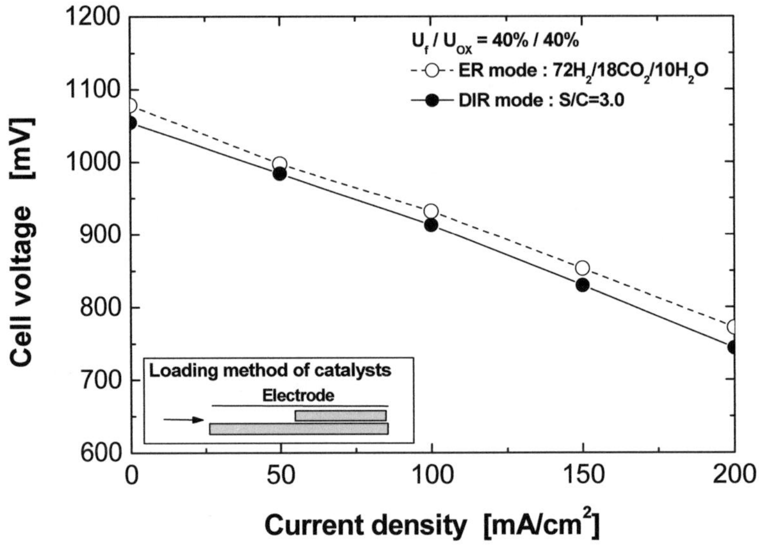

Fig. 10. Initial I-V performance of 250cm² DIR-MCFC single cell

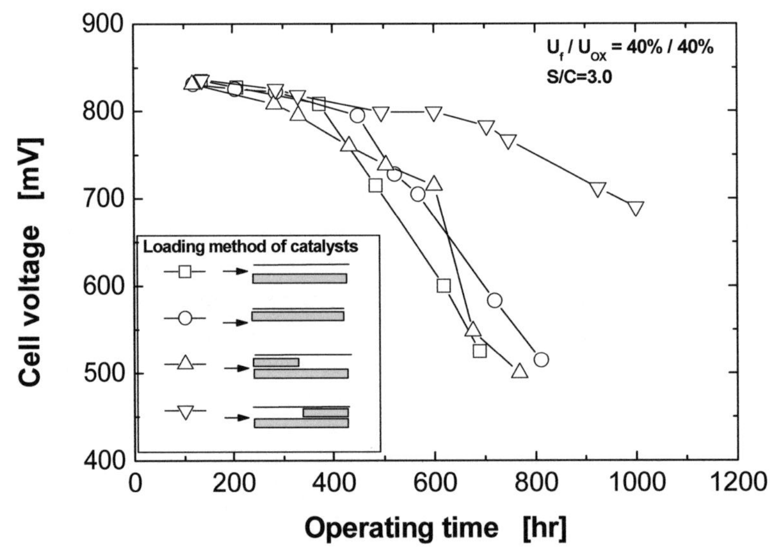

Fig. 11. Influence of the loading method of the reforming catalyst on the life performance

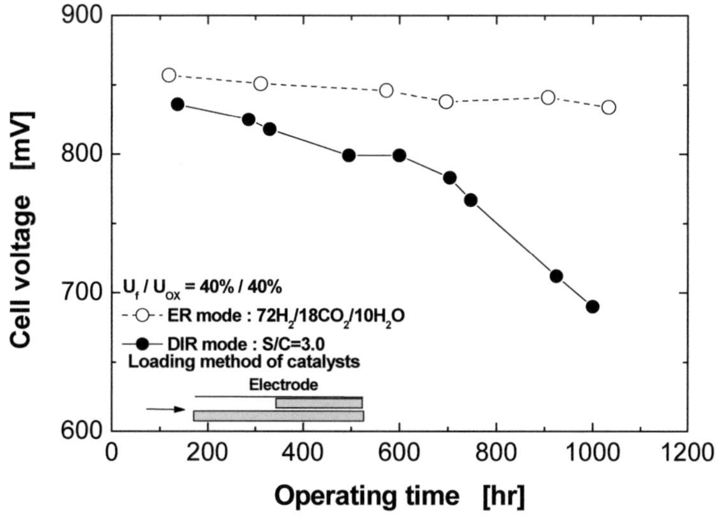

Fig. 12. Comparison of the cell performance between ER mode and DIR mode

Figure 13 shows comparison of the gas composition change in the cell according to operating time between "Upper stream" and "Downstream". Here, these results are that gases that collected at OCV (because the cell reaction does not occur, the ability of the catalyst is purely evaluated.) are analyzed by gas chromatograph. From the result of Fig.12 (a), because the hydrogen concentration of "Upper stream" is almost corresponding to the chemical equilibrium, the ability of the reforming catalyst is good. However, the hydrogen

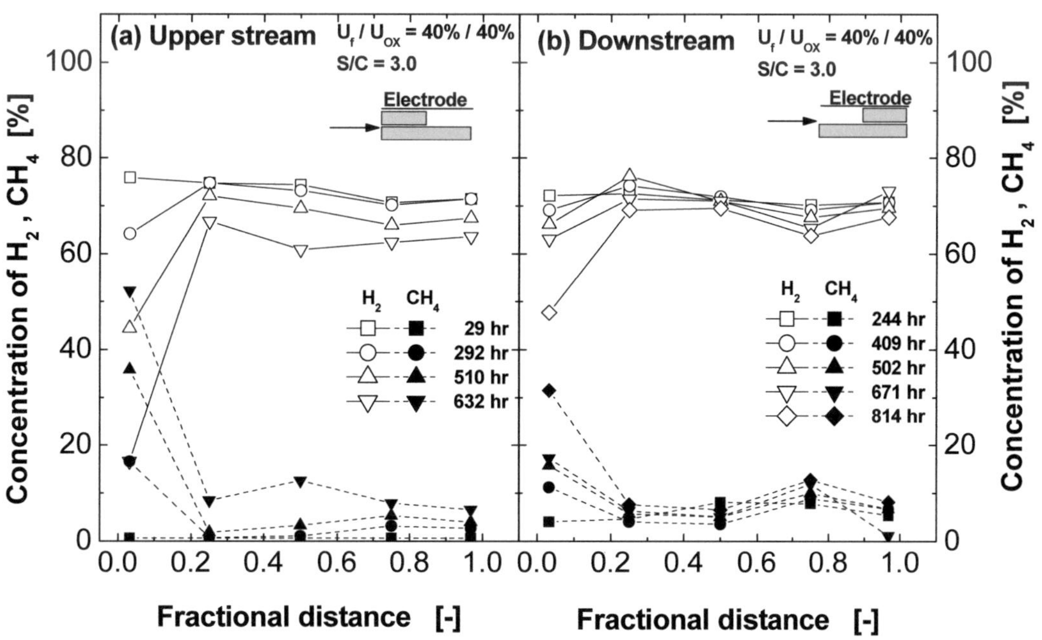

Fig. 13. Gas composition change in the cell according to operating time

concentration decreases and the methane concentration increases near the inlet when the operating time exceeds 292 hours. These concentration changes become large and progress toward the downstream side at 632 hours of the operating time. Therefore, the cell performance of "Upper stream" deteriorates for which the necessary hydrogen content cannot be secured by the catalyst pollution.

On the other hand, although the hydrogen concentration decreases and the methane concentration of "Downstream" increases near the inlet with the lapse of time, they are corresponding to the chemical equilibrium in 80% region of the entire cell. Moreover, they change rapidly at 814 hours of the operating time, and this change is corresponding to the drastic descent of the cell voltage. Therefore, the deterioration factor of the cell of "Downstream" is the pollution of catalyst loaded into the inlet as well as the cell of "Upper stream". Although the electrode of the downstream can use the hydrogen not used for the cell reaction of the hydrogen reformed in the upper stream part, the electrode of the upper stream cannot use the hydrogen if the catalyst of the upper stream cannot make the hydrogen by the catalyst pollution. Consequently, because the electrode of the upper stream hardly works when the operating time exceeds about 600 hours, the effective reaction area of the electrode decreases and the DIR-MCFC cannot generate electricity finally.

3.2 Analysis of the polluted catalyst

These four cells are disassembled to confirm the deterioration factor of DIR-MCFC, and are observed. Moreover, the polluted catalyst collected from the cell after experiment ends is analyzed by ICP for which it confirm that the molten salts make the reforming catalyst pollute. Figure 14 shows the sketch drawing of anode in each cell disassembled. All anode electrodes have discoloured from the parabolic and the gray to the black toward the exits from the inlet. There was no difference of these compositional variations though these different colours of electrodes were analyzed by an X-ray structure analysis.

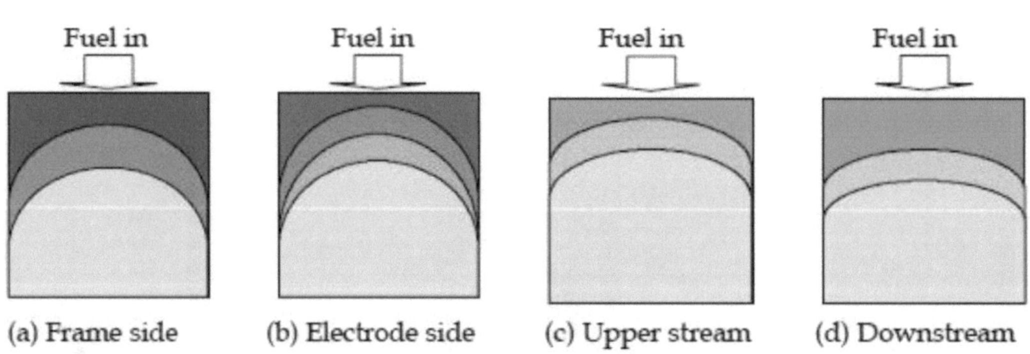

Fig. 14. Sketch drawing of anode in each cell disassembled

The reforming catalysts were collected according to this discoloration of each electrode after about 1000 hours of operating, and the amount of lithium and potassium in the catalyst were analyzed by ICP. Figure 15 shows the carbonate content in the collected catalyst from each cell. Here, amount of Li_2CO_3 and K_2CO_3 are converted from the results of lithium and potassium by ICP, and the location number means that it is a catalyst collected from the place shown in the number of the sketch of the electrode in each figure. Moreover, #4 and #5 of "Upper stream" and #6 and #7 of #Downstream" mean that the catalyst is collected from the electrode side of the corrugated current collector. The frame of electrode such as #1 of "Frame side", #5 of "Electrode side", #6 of "Upper stream" and #5, #7 of "Downstream" means that the catalyst is collected from near the wet seal part. Moreover, the colours of the catalysts collected from each location are corresponding to the colour of each electrode. Although the catalysts collected from the wet seal part and the upper stream part include much carbonate, the carbonate content of the catalysts collected from the centre of the cell and the downstream part are comparatively few. The carbonate content of the catalyst collected from "Downstream" is fewer than others. MCFC adopts the wet seal method that prevents the gas leakage by forming the liquid film with the molten salt between the matrix and the electrode. Therefore, a little molten salt begins to leak from the wet seal part into the cell. Especially, because the wet seal part of the upper stream is orthogonal to the gas flow, the leakage amount of the molten salts is much. However, because the wet seal part of the cell side is parallel to the gas flow, the leaked molten salt flows along the cell frame. The leaked molten salt doesn't, therefore, penetrate the catalytic layer easily. From these results, we understand that the discoloration of the parabolic in anode is signs where the electrolyte flows in the anode channel.

On the other hand, when the amount of lithium is compared with potassium, the amount of potassium is much. It can guess that potassium that is more than this composition is the molten salts of vapour phase because the molten salt of liquid phase is the same as the composition ratio of the electrolyte ($47Li_2CO_3/53K_2CO_3$; wt%). Therefore, Figure 15 was arranged to the ratio of the liquid phase molten salt and the vapour phase molten salt as shown in Fig.16. The liquid phase molten salt is more than the vapour phase molten salt in all location, and the vapour phase molten salt is distributed in the entire cell almost

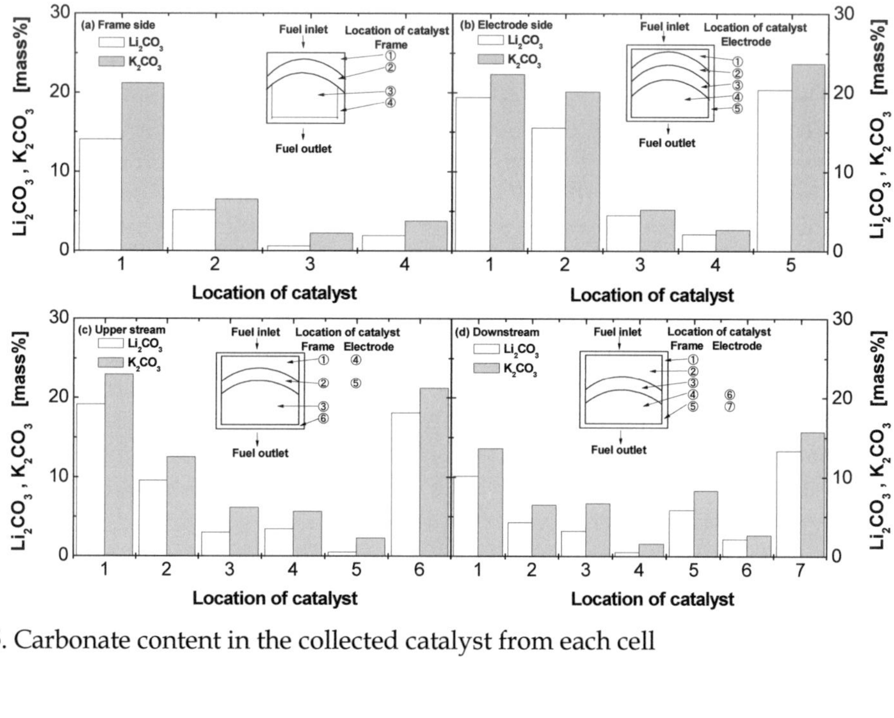

Fig. 15. Carbonate content in the collected catalyst from each cell

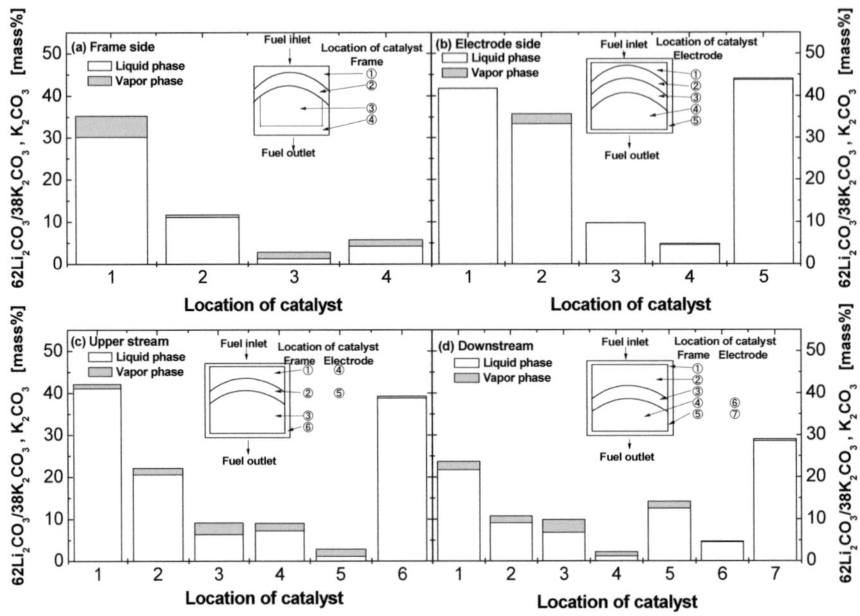

Fig. 16. Ratio of the liquid phase molten salt and the vapour phase molten salt in the polluted catalyst

uniformly. Generally, as for potassium, the deliquescence is stronger than lithium's, and the vapour partial pressure is also higher. Therefore, we elucidate the electrolyte volatile behaviour in MCFC using a visualization technique (Sugiura et al., 2008). The influence of electrolyte leakage from the wet seal section on the electrolyte volatile behaviour is evaluated by visualizing the wet seal. Here, because the Li/Na electrolyte is less volatile than the Li/K electrolyte, it is possible to pay attention to the electrolyte leakage from the wet seal part by excluding the influence of the volatizing phenomenon if $52Li_2CO_3/48Na_2CO_3$ electrolyte is used as the electrolyte. The volatilization phenomenon of the wet seal section is elucidated by observing a wet seal section via a visualization cell. Figure 17 shows the measurement image of the wet seal section in the anode gas channel immediately after operation. Figure 17 (a) shows the measurement image under OCV conditions, and Figure 17 (b) shows the measurement image under 125 mA/cm^2 of current density, respectively. Here, the electrode is supported only on one side because there is not the support for the cell frame side to observe the wet seal section. Therefore, this cell uses the current collector to prevent the electrode curving. From Fig.17 (a), we can see that volatile matter gushes from the wet seal section, which is a part of the anode and cell frame, and this gush is promoted by the increase of the current density, as shown in Fig.17 (b). This gush originates because the electrolyte of the wet seal section flows as a mist, with steam and CO_2 generated in the anode cell reaction. Moreover, this mist electrolyte volatilizes in the gas, is changed into a powder of Na_2CO_3 by reacting with CO_2, and it furthermore becomes a smaller mist by the gas flow. We understand that the gap of the wet seal part grows when the elasticity of the corrugated current collector decreases by the endothermic reaction of the catalyst, and the leakage of the molten salt from this grown gap is promoted, and the reforming catalyst is polluted by this leaked molten salt as shown in Fig.18. The catalytic activity of the reforming catalyst decreases by which the specific surface area

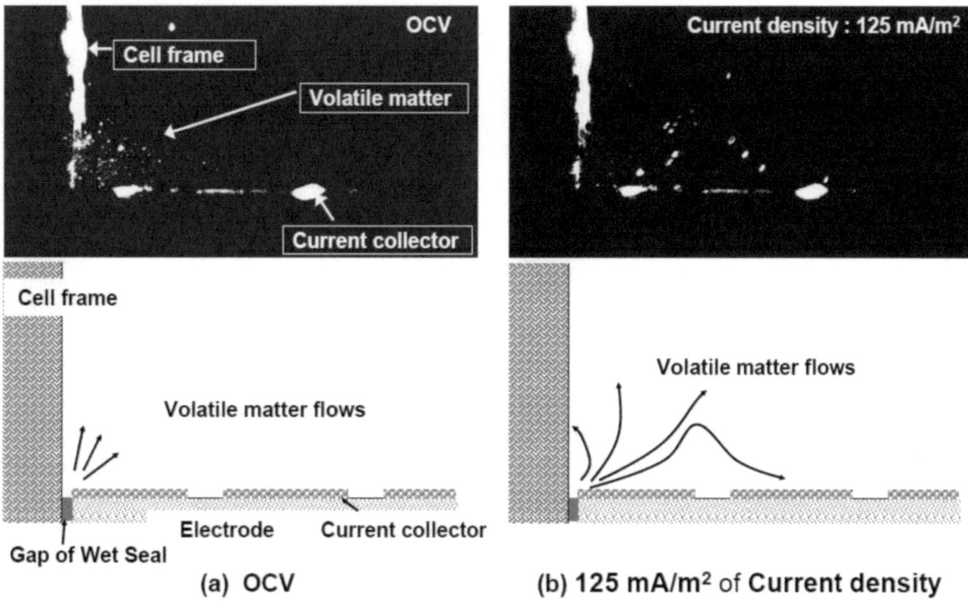

Fig. 17. Measurement image of wet seal part in the anode gas channel

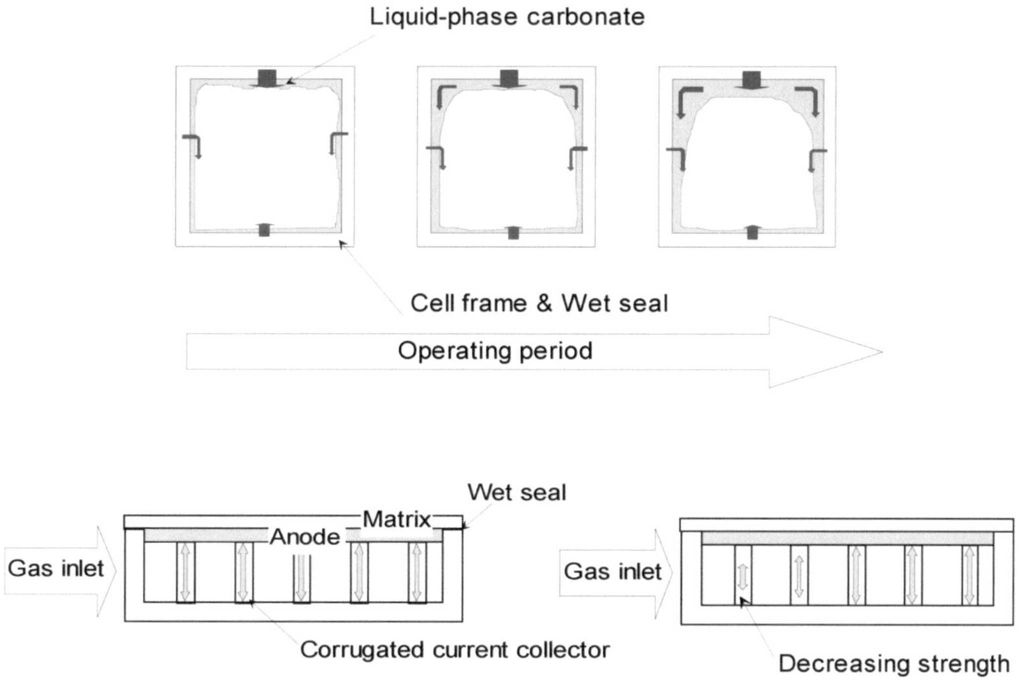

Fig. 18. Mechanism of the liquid phase pollution of reforming catalyst

decrease by adhering the leaked electrolyte to the reforming catalyst. The electrolyte that adheres to the reforming catalyst cannot be easily removed because the reforming catalyst has a micro pore structure (Sugiura & Ohtake, 1995). Moreover, the potassium in the leaked molten salt easily changes into KOH with the steam produced by the cell reaction, and disperses with the produced gas in the catalytic layer. Therefore, the catalyst pollution is divided into the liquid phase pollution and the vapour phase pollution. Though it is difficult to defend the catalyst from the vapour phase pollution, the catalyst can be defended from the liquid phase pollution for which the catalyst be covered by the material with large contact angle with the molten salt.

3.3 Prevention method of the liquid phase pollution

The corrugated current collector is plated by nickel which has large contact angle with molten salt to defend the catalyst from the liquid phase pollution. Figure 19 shows the effect of Ni plating on the cell performance. Here, the operating condition is DIR mode (U_f/U_{ox}=75%/50%; S/C=3.0), and the catalyst loading method adopted "Downstream" whose performance was the best in a current experiment. Moreover, Ni plating was applied to the corrugated current collector by the electrolytic process, and the thickness of Ni plating was about 100 microns. It is clear that the performance of the cell with the corrugated current collector with Ni plating is better than that of non plating. Moreover, the catalyst is collected after about 1000 hours of operating, and the carbonate content of the collected catalyst is analysed.

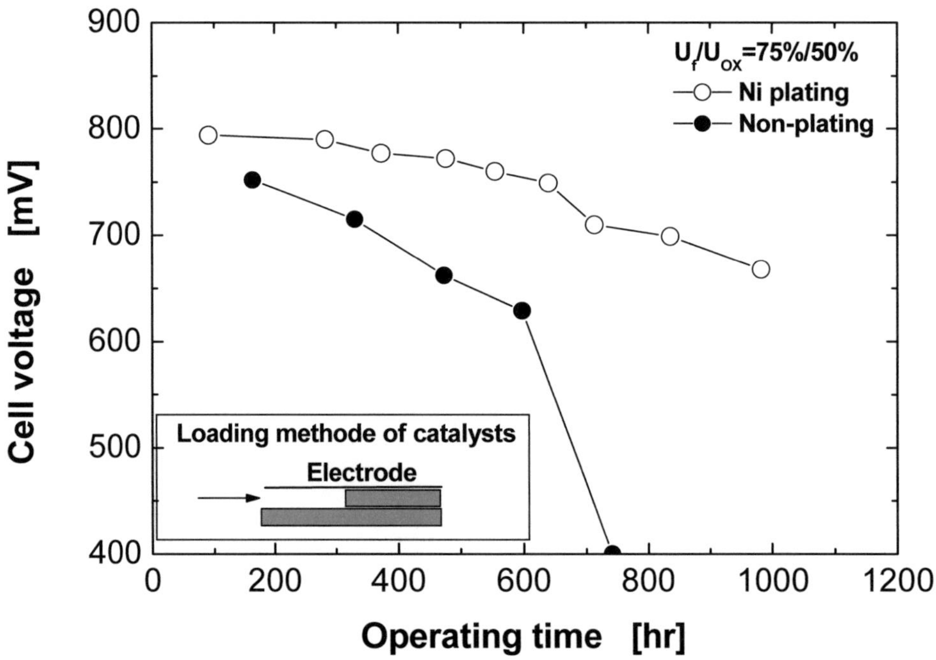

Fig. 19. Effect of Ni plating on the cell performance

Figure 20 shows the carbonate contents and the ratio of the liquid phase molten salt and the vapour phase molten salt in the polluted catalyst. Here, (a) and (b) are the carbonate content n the polluted catalyst, and (c) and (d) are the ratio of the liquid phase molten salt and the vapour phase molten salt in the polluted catalyst, respectively. It is clear that the carbonate content of the catalyst of the cell frame side by Ni plating decreases from that of non-plating. Therefore, the effect of Ni plating on the liquid phase pollution is able to be confirmed. However, the carbonate content of the catalyst of the electrode side has hardly changed though Ni plating was applied to the corrugated current collector. This reason is that because the catalyst of the electrode side is faced with the electrode, the path for the molten salt to reach the catalyst layer is different from the catalyst of the frame side, and there is a possibility that the catalyst contacts to the electrode. Therefore, the effect of Ni plating on the liquid phase pollution is small. From these results, it is clear that the liquid phase pollution can be controlled by applying Ni plating to the corrugated current collector.

On the other hand, the ratio of the vapour phase pollution of most catalysts in the cell is higher than that of the liquid phase pollution by controlling the liquid phase pollution. Because the vapour phase pollution is caused in the entire cell, the control of the vapour phase pollution is not easy like the liquid phase pollution. However, we elucidated that the vapour phase pollution originates in volatilizing of the molten salt that leaks from the wet seal part. Moreover, because this molten salt leakage from the wet seal part stops within one week after the operating temperature reaches, we understand that the catalyst pollution does not occur if this molten salt leakage can be controlled (Sugiura et al., 2010).

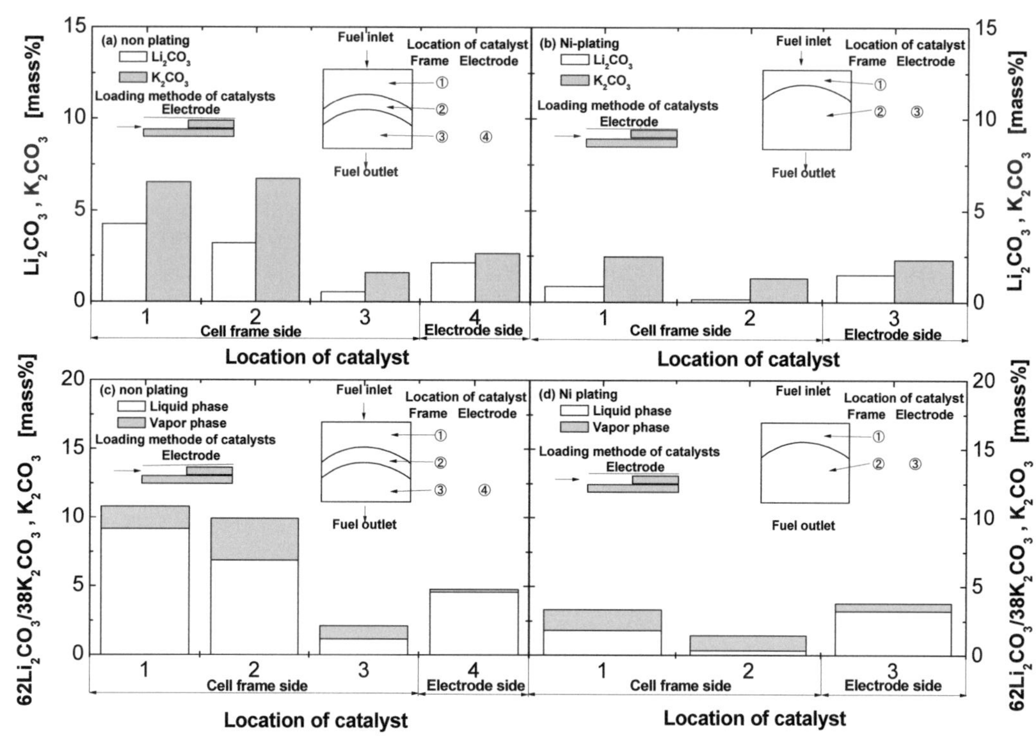

Fig. 20. Effect of Ni plating on the pollution ratio

4. Conclusion

The catalyst pollution in DIR-MCFC was elucidated from the cell test. The gas chromatograph is profitable for the evaluation of DIR-MCFC. The degradation factor of DIR-MCFC was the catalyst pollution by electrolyte, and the catalyst pollution is classified into the liquid phase pollution and the vapour phase pollution. The liquid phase pollution is the leaked electrolyte from the wet seal part, and the leaked electrolyte flows with the gas supplied along the cell frame and into the catalyst loading area. The gap of the wet seal part grows when the elasticity of the corugated current collector decreases by the endothermic reaction of the catalyst, and the leakage of the molten salt from this grown gap is promoted, and the reforming catalyst is polluted by this leaked molten salt. The catalystic activity of the refrming catalyst decreases by which the specific surface area decrease by adhering the leaked electrolyte to the reforming catalyst. The electrolyte that adheres to the reforming catalyst cannot be easily removed because the reforming catalyst has a micro pore structure. Therefore, because the liquid phase pollution has to be controlled physically, the catalyst was covered by the nickel with large contact angle with the molten salt. The performance of DIR-MCFC was improved using the corrugated current collector with Ni plating, which repels molten salts.

From the visualization experiment, because the electrolyte that polluted the reforming catalyst dispersed from wet seal section, DIR-MCFC can operate for long time if the reforming catalyst was protected from the electrolyte gush for 7 days from the operation-start. Moreover, though the most of the volatile matter of anode side was the electrolyte that gushed from the wet seal part, the part of volatile matter was KOH generated as the water generated by the cell reaction reacts with the electrolyte, and as the generated KOH returns to K_2CO_3 in the region where the concentration rose by the CO_2 generated by the cell reaction.

5. References

Miyake, Y.; Nakanishi, N.; Nakajima, T.; Itoh, Y.; Saitoh, T.; Saiai, A. & Yanaru, H. (1995). A study on degradation phenomena of reforming catalyst in DIR-MCFC, *J. Chem. Eng. Jpn.* Vol.21 (6), pp.1104–1109

Sugiura, K. & Ohtake, K. (1995). Deterioration of a catalyst's activity by Liquid-phase MC poisoning in DIR-MCFC, *J. Chem. Eng. Jpn.* Vol.21 (6), pp.1170–1178

Sugiura, K.; Naruse, I. & Ohtake, K. (1999) Deterioration of a catalyst's activity in direct internal reformingmolten carbonate fuel cells (effect of adsorption of vapour-phase molten carbonate), *J. Jpn. Soc. Mech. Eng.*, Ser.B 65 (629), pp330-336.

Sugiura, K.; Niwata, A.; Yamauchi, M. & Tanimoto, K. (2010). Influence of the segmented electrode use on electrolyte leakage in molten carbonate fuel cell, *ECS Transactions - 2009 Fuel Cell Seminar & Exposition, Volume 26,* pp.385-390

Sugiura, K.; Soga, M.; Yamauchi, M. & Tanimoto, K. (2008). Volatilization behaviour of Li/Na carbonate as an electrolyte in MCFC, *ECS Transactions - 2007 Fuel Cell Seminar & Exposition, Volume 12, issue 1,* pp.355-361

Tanimoto, K.; Yanagida, M.; Kojima,T.; Tamiya, Y.; Matsumoto, H. & Miyazaki, Y. (1998) Lomg-term operation of small-sized single molten carbonate fuel cells, *J. Power sources,* 72, pp77-82.

5

Recent Strategies in Organic Reactions Catalyzed by Phase Transfer Catalysts and Analyzed by Gas Chromatography

P. A. Vivekanand and Maw-Ling Wang*
Department of Environmental Engineering, Safety and Health, Hungkuang University, Shalu District, Taichung, Taiwan

1. Introduction

The field of catalysis provides chemists with new and powerful tools for the efficient synthesis of complex organic molecules. It is suitable for carrying out chemical and biochemical reactions at a high rate in nature as well as in organic synthesis. Catalysis expert's foresee catalysis among the most promising fields of basic research. This methodology can be applied to solve many fundamental, technological, environmental and social problems that face humanity. It finds application in modern chemical and petrochemical industries. Reactions involving two substances located in different phases of a reaction mixture are often inhibited due to the reagents inability to come into contact with each other. In such reactions, conventional techniques are environmentally and industrially unattractive. Nevertheless these reactions can be successfully promoted by a popular catalysis methodology *viz.*, phase-transfer catalysis (PTC) under mild operating conditions (Sasson & Neumann,1997; Mahdavi & Tamami, 2005; Barbasiewicz et al., 2006; Sharma et al., 2006; Devulapelli & Weng, 2009; Yang & Peng, 2010). Phase-transfer catalysts are being combined with enzymes in biotechnological processes and with transition metals in supramolecular chemistry and nano technology (Lancaster, 2002).

Currently, this key green approach (Makosza, 2000) is a powerful tool in the manufacture of fine chemicals and pharmaceuticals (Yadav & Bisht, 2004; Yadav & Badure2008). It has been recognized as a convenient and highly useful synthetic tool in both academia and industry because of several advantages of PTC *viz.*, operational simplicity, mild reaction conditions with aqueous media, suitability for largescale reactions, etc., which meet the current requirement of environmental consciousness for practical organic synthesis. This technique aids in the transfer of an ionic species from either an aqueous or a solid phase into the organic phase where the chemical reaction takes place. Undoubtedly, PTC offers many substantial advantages for the practical execution of numerous reactions (Wang & Tseng, 2002; Yadav, 2004; Wang & Lee, 2007; Wang & Lee, 2006; Wang et. al., 2003; Vivekanand & Balakrishnan, 2009a, 2009b, 2009c, 2009d, 2009e).

* Corresponding Author

Hence, PTC is now well recognized as an invaluable methodology for organic synthesis from two or more immiscible reactants and its scope and application are the subjects of current research. Consequently, there is an upsurge of interest in the synthesis of many more such catalysts which would be highly advantageous and may be employed as a vital tool in organic syntheses. Phase transfer catalysis will be of curiosity to anyone working in academia and industry that needs an up-to-date critical analysis and summary of catalysis research and applications. Presently, ingenious new analytical and process experimental techniques viz., ultrasound and microwave irradiation assisted PTC transformations (Masuno et al., 2005; Wang & Rajendran, 2006; Wang & Rajendran, 2007a, 2007b; Wang & Chen, 2008;. Wang & Chen, 2010; Vivekanand & Wang, 2011; Yang & Lin, 2011; Chatti, et al., 2002; Gumaste et al., 2004;. Luo et al., 2004;. Chatti et al., 2004; Bogdal et al., 2005; Hejchman et al., 2008; Baelen et al., 2008; Sahu et al., 2009; Awasthi et al., 2009; Greiner et al., 2009; Wang & Prasad, 2010; Fiamegos et. al., 2010) have become immensely popular in promoting various organic reactions.

Chemical kinetics is the study of the reaction rates of chemical reactions taking into account their reaction mechanism. Chemical kinetics is the basis of catalysis; however, catalysis is not a part of the kinetics. Mastering these reaction rates has many practical applications, for instance in understanding the complex dynamics of the atmosphere, in understanding the intricate interplay of the chemical reactions that are the basis of life and in designing an industrial process. Moreover, knowledge of kinetics will be helpful in developing theories that can be used to predict the outcome and rate of reactions.

Now a days, kinetic investigation of heterogeneous catalytic reactions is an indispensable step of the theoretical and applied investigations on catalysis. It facilitates in elucidation of the mechanism of a given heterogeneous catalytic reaction and contributes essentially to the revelation of the catalyst behavior in the course of its synthesis, utilization and recycles. Their examination assists in modeling and selection of optimal catalysts and optimization of catalytic reactors. Consequently, the growth of theory and practice in catalysis is implausible without unfolding extensive kinetic investigations.

Catalytic reactions and reaction kinetics are usually monitored using well-established analytical techniques such as gas chromatography (GC). For determining the kinetics of any reaction, samples were collected from organic layer at regular intervals, diluted with suitable solvents and finally injected into GC for analysis. Retention time and area of reactants were obtained from the chromatograph. Using the obtained data's, rate constants were evaluated from the kinetic plot. Thus, gas chromatography identifies compounds by chromatography retention time and thereby, making the method a highly accurate procedure and an essential method in the analysis of organic reactions. The analysis of compounds by GC is very fast, accurate and reliable. Further small samples can be analyzed with high resolution.

Anionic compounds, including anions of organophosphoric acids, carboxylic acids and phenols in aqueous samples can be directly determined by liquid chromatography (LC) or ion chromatography. Alternatively, gas chromatography (GC) analysis after extraction and derivatization is a very practical option. Generally, these types of GC procedures are based on the isolation of anionic analytes from the aqueous samples followed by purification, desiccation, derivatization and analysis by GC (Mikia, et al., 1997). Capillary electrophoresis and ion chromatography identify inorganic anions by retention time or migration time. Nevertheless, a sample contaminated with matrix is often difficult to analyze.

Analysis of inorganic anions in foresenic chemistry (Sakayanagi, et al., 2006), iodide anion(Lin, et al., 2003), flavonoids(Yiannis, et al., 2004), haloacetic acids in drinking waters (Cardador, et al., 2008), phenolic metabolites in human urine(Bravo, et al., 2005), phenols(Fiamegos, et al., 2008) etc., have been carried out using GC under PTC conditions.

In view of the success and vitality of GC analysis in PTC assisted organic reactions, we have proposed to present recent happenings in the field of PTC and to study its applications to various organic reactions that were monitored by gas chromatography. Further, kinetics of various organic reactions catalyzed by PTC carried out under a wide range of experimental conditions will be presented. Much of our current effort is devoted to exploring the kinetic aspects of the reactions, the role of the catalyst structure, and influence of experimental parameters. The roles of these species in the overall catalytic organic reactions are investigated so as to understand whether they are mere spectators or participate in the reaction and how their presence affects the overall reaction kinetics.

Divided into five sections, the chapter explores, Suzuki-coupling, epoxidation, C-alkylaion, N-alkylation and O-alkylation reactions. Helping readers to better understand the kinetics of the PTC reactions that are analyzed by GC, the examples in the chapter substantiates the development of more effective PTC processes achieved during the last few decades, enabling industry to embark on a safer and more efficient synthesis of organic compounds for the manufacture of a wide array of products.

2. Phase transfer catalyst assisted organic reactions followed by gas chromatography

2.1 N-alkylation

Imide derivatives are organic compounds with numerous applications in biology (Langmuir et.al., 1995; Settimo et.al., 1996)as well as in synthetic (Ohkubo, et al. 1996) and polymer chemistry(Iijima, et al., 1995). GC analysis of PTC assisted *N*-alkylation reactions are well documented (Jankovic, et al., 2002; Mijin, et.al., 2004 & Mijin, et.al., 2008). Synthesis of *N*-butylphthalimide (PTR) can be achieved by reacting 1-bromobutane and potassium salt of phthalimide in a liquid (water)–liquid (organic solvent) two-phase medium catalyzed by quaternary ammonium salt. Nonetheless, the hydration of potassium salt of phthalimide in aqueous solution is serious which results in poor yield.

Herein, we discuss the kinetics of synthesis of *N*-butylphthalimide (PTR) (Scheme 1) that was analyzed by GC (Wang, et al., 2005). The kinetic experiments were run in an ordinary smooth-wall, three-necked flask, fitted with an agitator, reflux condenser and sampling port. All ingredients *viz.*, potassium phthalimide(excess agent), TBAB and acetonitrile, were placed in the flask and stirred at 800 rpm for about 30 min at 70 ^{0}C (Wang, et al., 2005). Measured quantities of *n*-bromobutane (limiting agent) and toluene (internal standard) were then added to reactor. The reaction mixture was stirred at 800 rpm. Samples of the organic phase were withdrawn at regular time intervals by stopping the stirrer for 15-30 sec such that the organic phase had separated well enough to get a good sample, then analyzed by GC using an internal standard method. The conditions for the GC analysis are as follows: Shimadzu GC 17A, J&W

Scientific Inc., capillary column (db-1 column); 100% poly(dimethylsiloxane) stationary phase; 15m x 0.525m column dimension; carrier gas, nitrogen (60 ml/min); flame ionization detector; injection temperature: 250 ⁰C (Fig. 1). An aliquot of reaction mixture (0.5 ml) was injected and the retention times for acetonitrile, 1-bromobutane, toluene and *N*-butylphthalimide are presented in Table 1. The structure of the product **3** was confirmed through GC mass spectroscopy(Fig. 2), which showed a molecular ion peak at 203 (M^+).

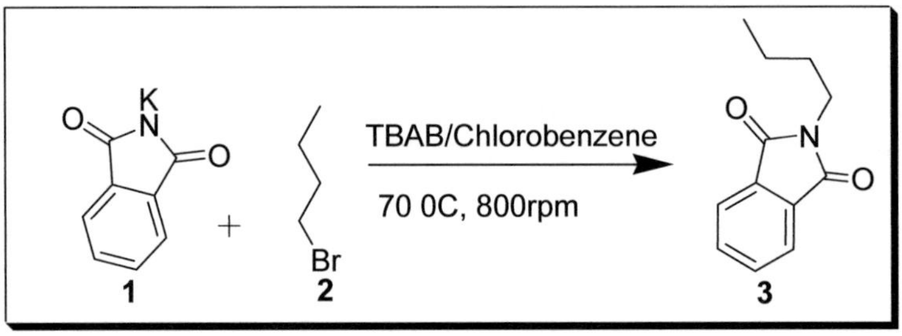

Scheme 1. *N*-alkylation of potassium salt of phthalimide catalyzed by TBAB (Wang, et al., 2005).

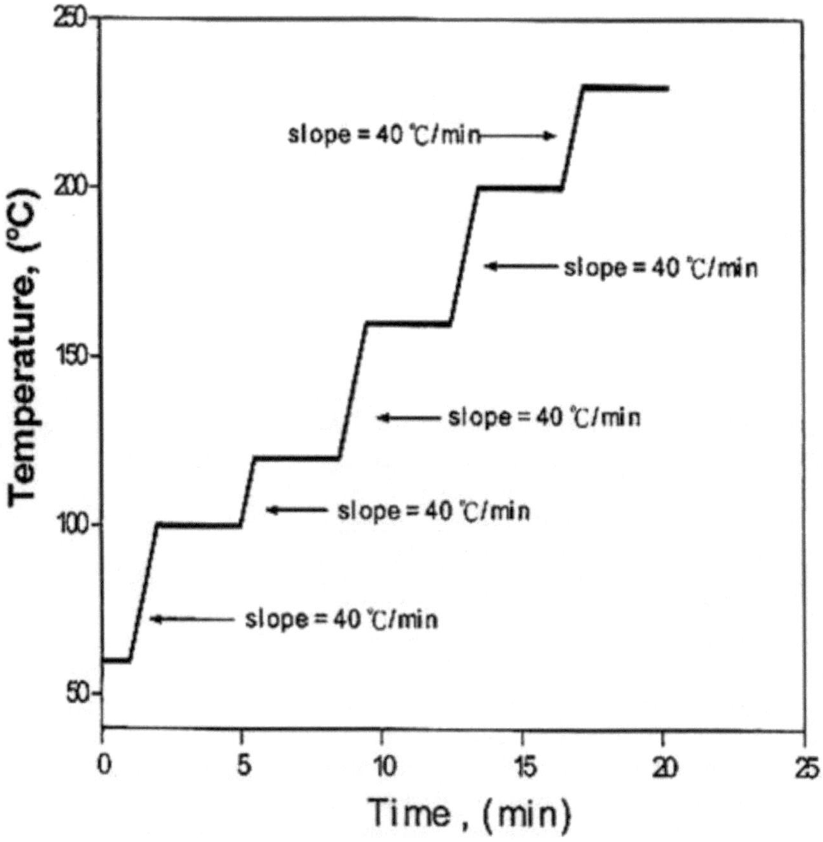

Fig. 1. GC analysis (temperature programming) condition for following N-Alkylation of potassium salt of phthalimide

Entry No	Solvent/Reactant/ Internal Standard/Product	Retention Time (Min.)
1	Acetonitrile	1.42
2	1-Bromobutane	3.03
3	Toluene	3.50
4	*N*-(*n*-butyl)phthalimide	12.79

Table 1. Determination of retention time for compounds in the N-alkylation reaction mixture by GC analysis

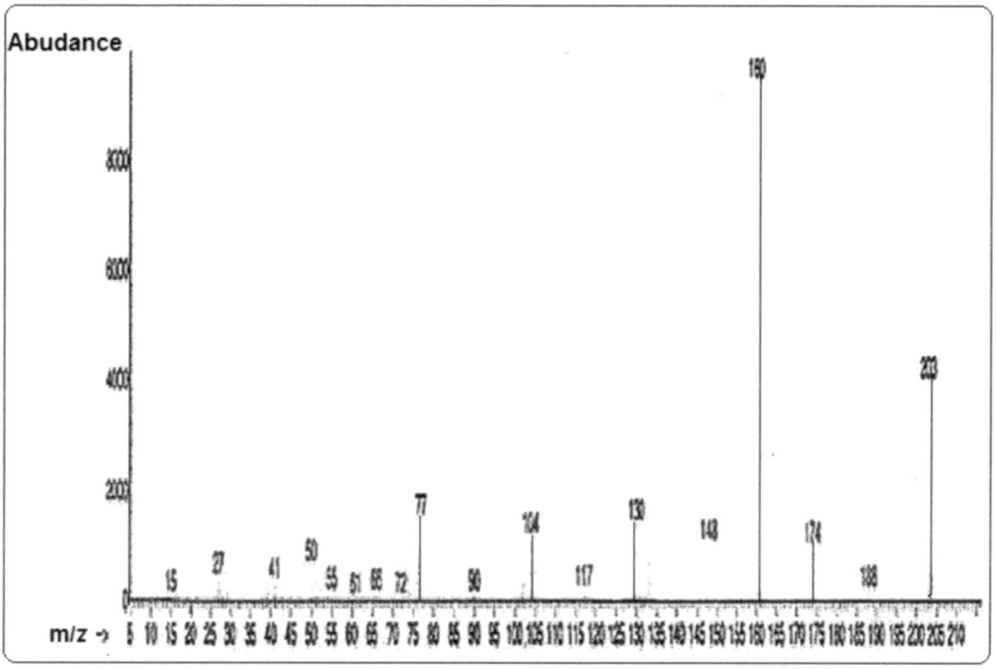

Fig. 2. GC-MS analysis of *N*-(*n*-butyl)phthalimide

2.1.1 Reaction mechanism and kinetic model

PTC assisted *N*-alkylation of potassium salt of phthalimide (PTK) with organic substrate was carried out in a solid–liquid solution (organic solvent) under phase-transfer catalysis conditions (Scheme 1).The reaction was carried out in the absence of water so as to avoid hydration. On comparing the reaction under liquid (organics solvent)-liquid (water) phase-transfer catalysis conditions (LL-PTC) with solid–liquid solution (organic solvent) under phase-transfer catalysis conditions (SL-PTC), the later method is advantageous because reaction rate is greatly enhanced and the yield of the product is increased. In the beginning of the reaction, the active catalyst *N*-(tetrabutylammonium) phthalimide (PTQ) is produced from the reaction of PTK, dissolved in organic solvent, with catalyst QBr. The inorganic salt KBr precipitated as a solid form from the organic-phase solution. Then, PTQ reacts with organic-phase reactant RX to produce the desired product PTR (Scheme 2).

$$PTK_{(s)} + QX_{(org)} \xrightarrow{\text{Organic solvent}} PTK_{(org)} + QX_{(org)} \tag{1}$$

$$PTK_{(org)} + QX_{(org)} \xleftrightarrow{K_1} PTQ_{(org)} + KX_{(org)} \tag{2}$$

$$PTQ_{(org)} + KX_{(org)} \xleftrightarrow{K_2} PTQ_{(org)} + KX_{(s)} \tag{3}$$

$$PTQ_{(org)} + RX_{(org)} \xrightarrow{\text{Organic solvent}} PTR_{(org)} + QX_{(org)} \tag{4}$$

Scheme 2. Mechanism for the *N*-alkylation of potassium salt of phthalimide catalyzed by TBAB.

In the Scheme 2, $QX_{(org)}$ and $RX_{(org)}$ represent the quaternary ammonium salt in the organic-phase solution and organic-phase reactant (*n*-bromobutane), respectively. The subscripts "org" and "s" denote the species in organic-phase and in solid-phase, respectively. The overall reaction is expressed as:

$$PTK_{(s)} + RX_{(org)} \xrightarrow{\text{Organic solvent, QX}} PTR_{(org)} + KX_{(s)} \tag{5}$$

As shown in eq 2, the reaction is fast and reaches equilibrium in a short time. Thus, the equilibrium constant K_1 is defined as:

$$K_1 = \frac{[KX]_{org}[PTQ]_{org}}{[PTQ]_{org}[QX]_{org}} \tag{6}$$

As stated, the inorganic salt KX precipitates from the organic solution and the equilibrium constant K_2 is defined as:

$$K_2 = \frac{[KX]_s}{[KX]_{org}} \tag{7}$$

The rate equation of the intrinsic reaction is given in Eq. (4):

$$\frac{d[PTR]_{org}}{dt} = -\frac{d[RX]_{org}}{dt} = k_{\text{int}}[PTQ]_{org}[RX]_{org} \tag{8}$$

where k_{int} is the intrinsic rate constant. The material balance for the catalyst is given by:

$$[QX]_{org,i} = [QX]_{org} + [PTQ]_{org} \tag{9}$$

where $[QX]_{org,i}$ is the initial concentration of QX. On solving Eqs. (6), (7) and (9), we obtain:

$$[PTQ]_{org} = f_c[QX]_{org,i} \tag{10}$$

where f_c is given as

$$f_c = -\frac{1}{1 + \frac{1}{K_1 K_2}\frac{[KX]_s}{[PTK]_{org}}} \tag{11}$$

We assume that the concentrations of $[KX]_s$ and $[PT\text{-}K]_{org}$ are kept at constant values after the induction period of the reaction. Therefore, as shown in Eq. (10), $[PTQ]_{org}$ is kept at a constant value. For this, Eq. (7) can be expressed as:

$$-\frac{d[RX]_{org}}{dt} = k_{app}[RX]_{org} \tag{12}$$

where k_{app} is the apparent rate constant of the pseudo first-order rate law.

$$k_{app} = k_{\text{int}}[PTQ]_{\text{org}} = k_{\text{int}} f_c [QX]_{\text{org,i}} \tag{13}$$

Eq. (12) is integrated:

$$-\ln(1 - X) = k_{app}t \tag{14}$$

where X is the conversion of 1-bromobutane (RX), i.e.

$$X = 1 - \frac{[RX]_{org}}{[RX]_{org,i}} \tag{15}$$

where $[RX]_{org,i}$ is the initial concentration of RX in the organic-phase solution. From Eq. (14), it is obvious that the reaction follows a pseudo first-order rate law. By plotting $-\ln(1 - X)$ versus t, the apparent rate constant k_{app} is obtained experimentally from the slope of the straight line.

In general, the hydration of potassium salt of phthalimide in aqueous solution is a serious problem and meager yield of products were obtained when the reaction of potassium salt of phthalimide and organic substrate was carried out under liquid (organics solvent)–liquid (water) phase-transfer catalysis conditions (LL-PTC). Hence, in this work, phase-transfer catalysis was successfully employed to synthesize *N*-alkylphthalimide (PTR) from the reaction of potassium salt of phthalimide (PTK, as excess reagent) with alkylating agent (RBr, as limiting reagent) in solid–liquid phase-transfer catalysis conditions (SL-PTC). Under appropriate conditions, a high yield of the product was obtained. The product was successfully separated and purified from the solid–liquid phase reaction solution. The kinetic results show a material balance between reactants and products, i.e., the consumption of the amount of reactant (*n*-bromobutane) equals to the sum of the generation of the amount of the product (*N*-alkylphthalimide). From GC analysis (Fig.2 and Table 1) and kinetic results, no byproducts were observed during or after the reaction system, indicating that only PTR was produced from the reactant RBr by phase transfer catalysis conditions. Therefore, the consumption of the reactant equals the production of product. The kinetics results obtained from the plot of –ln(1-X) *vs.* time using GC analysis under various conditions are discussed in the following sections.

2.1.2 Influence of stirring speed

The effect of agitation speed on the rate of the reaction under standard reaction conditions were investigated by varying the agitation speed in the range of 0–1100 rpm. From the plot of -ln(1-X) vs. time, the apparent rate constants (k_{app}) were evaluated (Fig. 3). It is clear that the reaction follows the pseudo first-order rate law. The conversion is increased with the increase in agitation speed up to 400 rpm, but there is no significant improvement in the reaction by further increasing the agitation speed from 400 to 1,000 rpm. This phenomenon indicates the less influence of the external mass transfer resistance on the reaction beyond 200 rpm. Therefore, the agitation speed was set at 800 rpm for studying the reaction phenomena at which the resistance of mass transfer stays at a constant value. We observed similar trend in the kinetic study of synthesizing 1-(3-phenylpropyl)pyrrolidine-2,5-dione under solid-liquid phase-transfer catalytic conditions (Wang & Chen, 2008).

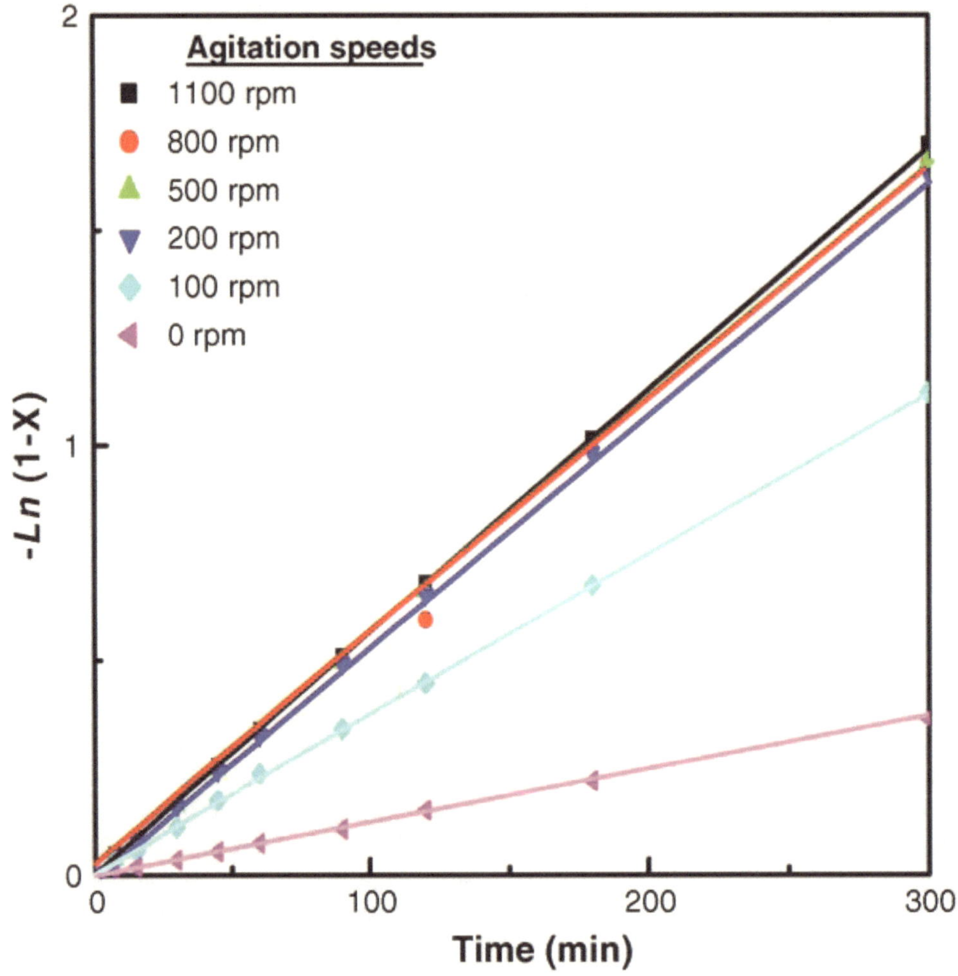

Fig. 3. Effect of stirring speed on the rate of *N*-alkylation of PTK: 6 mmol of PT-K, 0.7 mmol of TBAB, 50 mL of acetonitrile, 4 mmol of *n*-bromobutane, 0.5 g of toluene (internal standard), 70 ^{0}C. (Reprinted with permission from Wang, et al., 2005. Copyright (2011) Elsevier).

2.1.3 Effect of different catalysts

In this work, eight quaternary ammonium salts *viz.*, THAB, TBAB TOAB, BTEAB, QSO_3, THAB, TBAB and TOAB were used to examine their reactivity. In principle, there is no universal rule to guide in selecting an appropriate phase-transfer catalyst except that determined from experiments. The reason is that different reactions need various catalysts to enhance the rate and to promote the yields. From the plot -ln(1-*X*) *vs* time, the rate constants were obtained (Fig . 4). The order of the reactivities of these onium salts is: THAB> TBAB> TOAB> BTEAB > QSO_3. TEAB and QSO_3 (4-(trialkylammonium) propansultan), which are more hydrophilic, do not possess high reactivity. THAB, TBAB and TOAB of appropriate hydrophilic and hydrophobic properties exhibit high reactivity to obtain high conversion of 1-bromobutane. Further, it is favorable for the reaction in choosing quaternary ammonium salts of larger carbon numbers. The reason is that the lipophilicity is strong using the quaternary ammonium salts of larger carbon number(Wang & Rajendran, 2007b, 2007c; Wang & Chen, 2008).

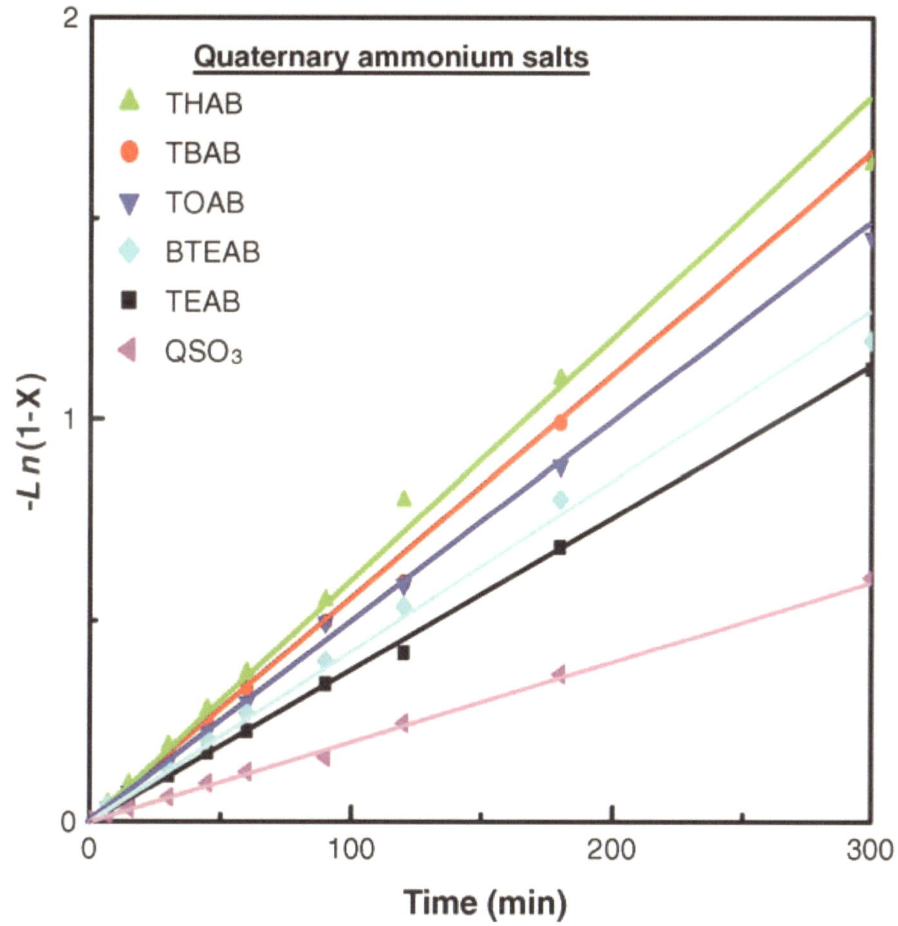

Fig. 4. A plot of -ln(1-X) of 1-bromobutane *vs.* time with different quaternary ammonium salts; 4 mmol of 1-bromobutane, 6 mmol of phthalimide potassium salt, 50 ml of acetonitrile, 0.7 mmol of PTC, 0.5 g of toluene (internal standard), 800 rpm, 70 °C. (Reprinted with permission from Wang, et al., 2005. Copyright (2011) Elsevier).

2.2 O-alkylation

In this work, ultrasonic irradiation assisted synthesis of dimethoxydiphenylmethane (DMODPM) from the reaction of methanol and dichlorodiphenylmethane (DCDPM) was successfully carried out in a liquid-liquid phase-transfer catalytic (LLPTC) reaction (Scheme 3) (Wang & Chen, 2009). Hydrolysis of the ketal product in acidic solution is avoided by carrying out the reaction in a basic solution. Two major advantages of carrying out the reaction under PTC conditions are i) it enhances the reaction and increases the yield and ii) also minimizes the by products.

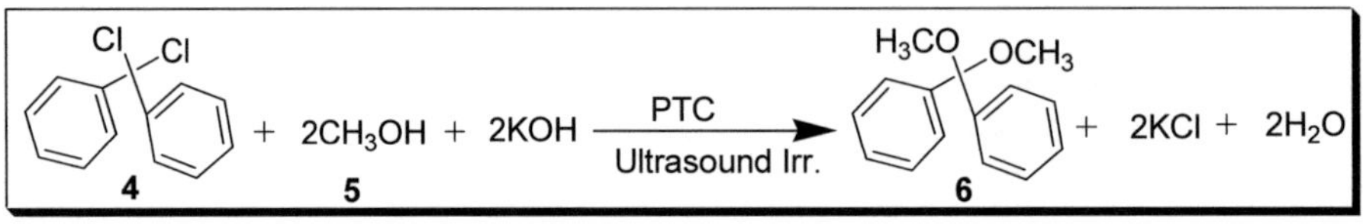

Scheme 3. Synthesis of DMODPM under LLPTC conditions(Wang & Chen, 2009).

The effects of the reaction conditions on the conversion of DCDPM, as well as the apparent rate constant ($k_{app,1}$) of the first reaction in the organic-phase solution, were investigated in detail. The product DMODPM and the reactants (DCDPM and ethanol) were all identified by GC-MS and NMR and IR spectroscopies. The GC mass spectrum of **6** showed a peak at

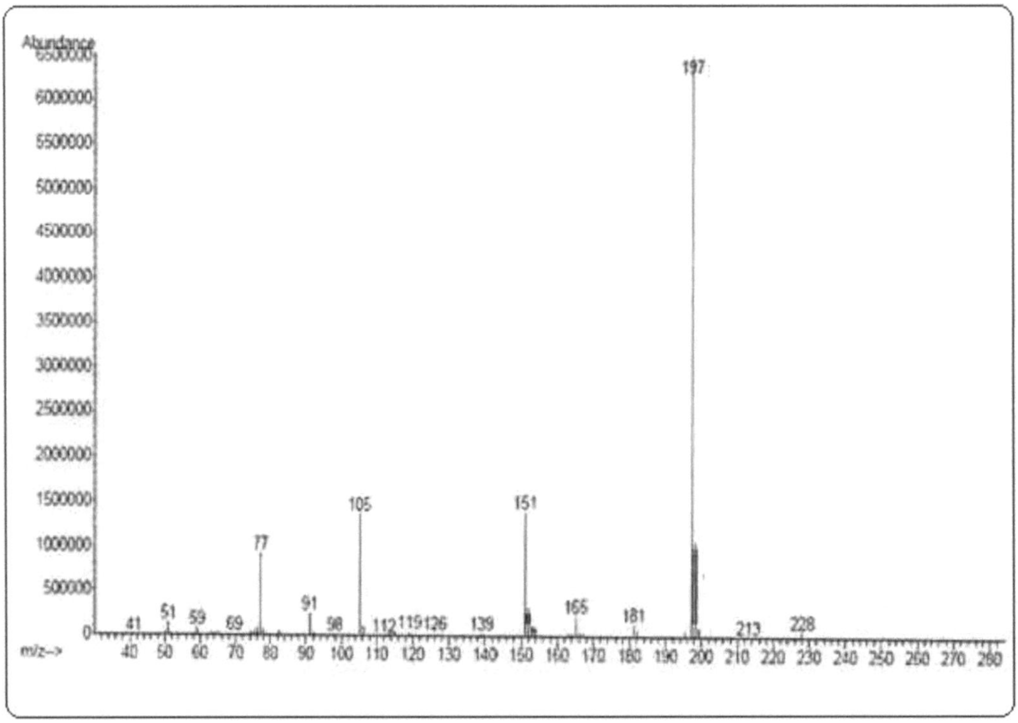

Fig. 5. GC-MS chromatogram of dimethoxydiphenylmethane(**6**).

m/z 228 (M^+) (Fig.5) Their concentrations (or contents) were analyzed by GC with GC17A model instrument (Shimadzu). The stationary phase was 100% poly(dimethylsiloxane). The carrier gas was N_2 (30 mL/min). The column was db-1 type.

2.2.1 Reaction mechanism and kinetic model

We believe that methanol first dissolves and reacts with KOH to produce potassium methoxide (CH_3OK or MeOK) in the aqueous solution. Then, CH_3OK further reacts with TBAB catalyst (QBr) to form tetrabutylammonium methanoxide (MeOQ or QOR), which is an active organic-soluble intermediate. This active intermediate (MeOQ) then reacts with DCDPM through two sequential reaction steps in the organic phase to produce the desired product, dimethoxydiphenylmethane (DMODPM). The reaction mechanism of the overall reaction is expressed in Scheme 4.

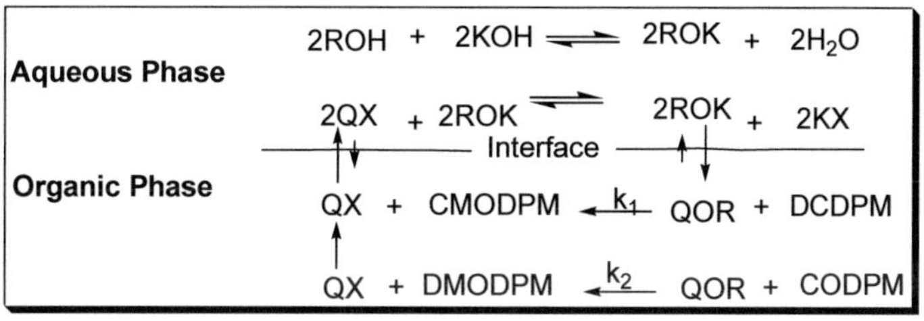

Scheme 4. Mechanism of DMODPM synthesis under LLPTC conditions.

where ROH, ROK, and QOR represent methanol, potassium methoxide, and tetrabutylammonium methanoxide, respectively; CMODPM and DMODPM are the monochloro-substituted (chloromethoxydiphenylmethane) and dichloro-substituted (dimethoxydiphenylmethane) products, respectively; and QX is the quaternary ammonium salt, where X can be either chloride or bromide. k_1 and k_2 are the two intrinsic rate constants of the organic-phase reactions.

For a two-phase phase-transfer catalytic reaction, the rate is usually determined by four steps, i.e., (a) the ionic aqueous-phase reaction, (b) the organic-phase reaction, (c) the mass transfer of species QOR (active intermediate) from the aqueous phase to the organic phase, and (d) the mass transfer of species of the regenerated catalyst QBr from the organic phase to the aqueous phase. The mass transfers of species from the aqueous phase to the organic phase and *vice versa* are all fast. The ionic aqueous-phase reaction is also very fast. Therefore, it is obvious that the organic-phase reaction, which is usually slow, is the rate-determining step. The ionic reaction in aqueous solution is fast. Also, MeOQ formed from the reaction of potassium methoxide and tetrabutylammonium bromide (TBAB) is an organic-soluble compound. The transfer of MeOQ from the aqueous phase to the organic phase is also fast. Therefore, the two sequential reactions in the organic phase are the rate-determining steps for the whole reaction. From the GC spectrum of the reaction samples, only DMODPM product was observed and no CMODPM was observed. This fact indicates that the second reaction is faster than the first one. Following the Bodenstein steady-state assumption, the production rate of CMODPM equals the consumption rate of CMODPM in the reaction solution. Once CMODPM is produced, it reacts with QOR very quickly to produce the final product DMODPM in the second reaction of the organic phase. Thus, the first reaction in the organic phase is the rate-determining step. Also, CMODPM was not observed. Thus, the rate of the change of CMODPM with respect to time was set to be zero, as shown in Eq 16. Consequently, the first reaction in the organic phase is the rate-determining step. Thus, we have

$$\frac{d[CMODPM]_{org}}{dt}=0 \tag{16}$$

where the subscript "org" denotes the species in the organic solution. The material balances for DCDPM, CMODPM, and DMODPM in the organic-phase solution are

$$-\frac{d[DCDPM]_{org}}{dt}=k_1[DCDPM]_{org}[QOR]_{org} \tag{17}$$

$$-\frac{d[CMODPM]_{org}}{dt}=k_1[DCDPM]_{org}[QOR]_{org}-k_2[CMODPM]_{org}[QOR]_{org} \tag{18}$$

$$\frac{d[DMODPM]_{org}}{dt}=k_2[QOR]_{org}[CMODPM]_{org} \tag{19}$$

Combining eqs. (16) and (18) results in,

$$[CMODPM]_{org}=\frac{k_1}{k_2}k_2[DCDPM]_{org} \tag{20}$$

From eqs. (17), (19) and (20), we get

$$-\frac{d[DCDPM]_{org}}{dt}=\frac{d[DMODPM]_{org}}{dt} \tag{21}$$

This result indicates that the consumption rate of DCDPM equals the production rate DMODPM in the organic phase. No other byproducts were observed during or after the reaction. Therefore, by integrating the equation after combining Eqs 17, 20, and 21, we have

$$-\ln(1-X)=k_{app}t \tag{22}$$

where $k_{app,1}$ is the apparent rate constant and X is the conversion of DCDPM, i.e.

$$X=\frac{[DMODPM]_{org}}{[DCDPM]_{org,i}}=\frac{[DCDPM]_{org,i}-[DCDPM]_{org}}{[DCDPM]_{org,i}} \tag{23}$$

$$k_{app,1}=k_1[QOR]_{\mathrm{org}} \tag{24}$$

where the subscript i represents the initial conditions of the species. The rate at which the reactant DCDPM is consumed can be calculated from Eq 17 and the rate at which the final product DMODPM is produced can be calculated from Eq 19. By applying the pseudo-steady-state approach, the rate of the final product (DCDPM) can be calculated from Eq 17.

As shown in Eq 22, it is obvious that the reaction follows a pseudo-first-order rate law. The $k_{app,1}$ values were obtained by plotting the experimental data for -ln(1 - X) *vs.* time (t). Thus, the reaction rate was calculated from Eq 17. Unexpected products were noticed in the absence of KOH and phase-transfer catalyst. However on the addition of KOH and PTC, DMODPM product was obtained in a small quantity. Therefore to enhance the reaction rate greatly, ultrasonic irradiation was employed in the reaction. The kinetics results obtained based on GC analysis (Fig. 6) under various conditions is discussed in the following sections.

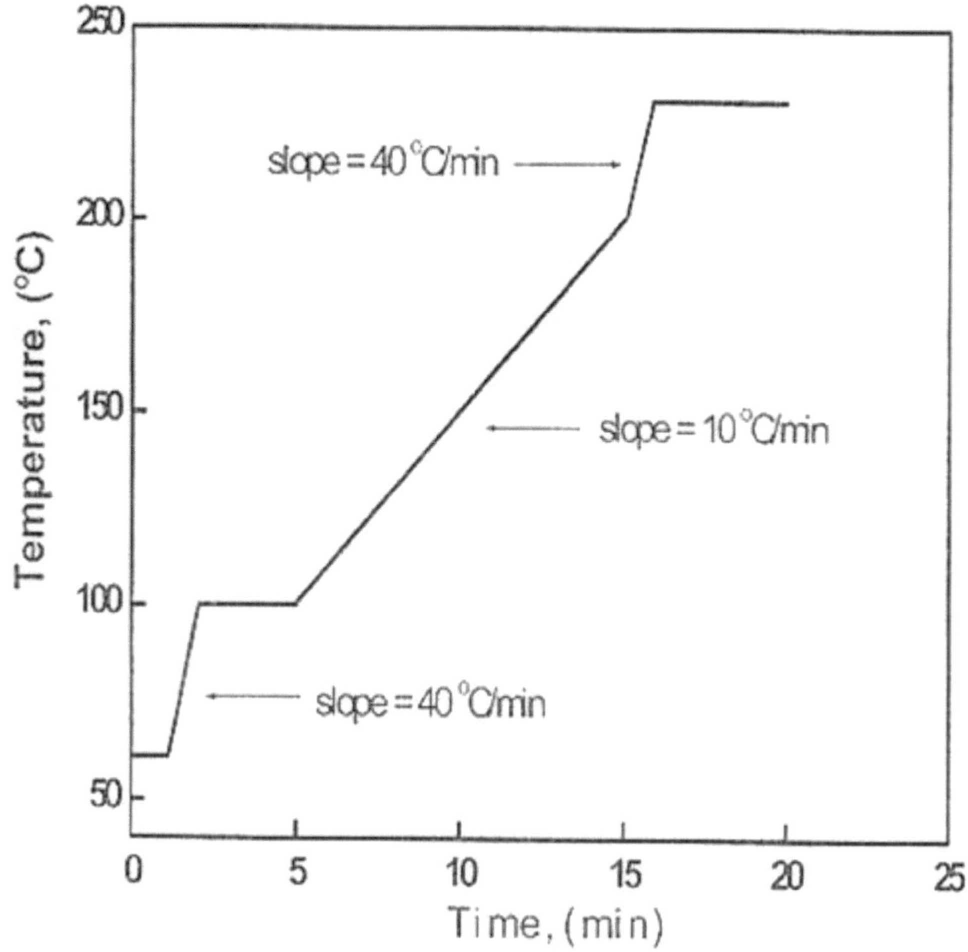

Fig. 6. GC analysis (temperature programming) condition for following O-alkylation

2.2.2 Effect of ultrasonic power and frequency

To ascertain the influence of various ultrasonic frequencies on the rate of the two phase reaction of DCDPM and methanol with same output power of 300 W, the ultrasonic frequency was varied in the range of 20-50 kHz under otherwise similar conditions using TBAB as the catalyst. The kinetic profile of the reaction is obtained by plotting -ln(1 - X) versus time (Fig. 7). From these observed results, it can be inferred that ultrasonic assisted phase-transfer catalysis significantly increases the rate of the reaction.

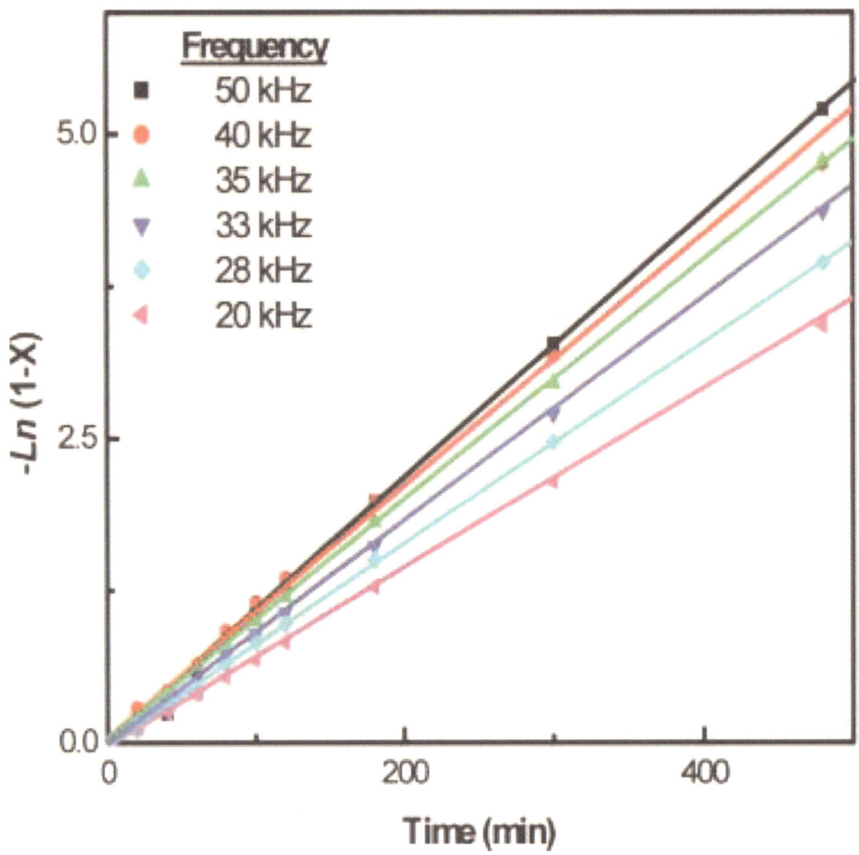

Fig. 7. Effect of the ultrasonic frequency on the conversion of DCDPM. Conditions: 0.228 g of TBAB, 90 mmol of methanol, 2.75 mmol of DCDPM, 40 mL of chlorobenzene, 5 g of KOH, 10 mL of water, 0.1 g of toluene, 400 rpm, 60 °C 300 W. (Reprinted with permission from Wang & Chen, 2009. Copyright (2011) American Chemical Society)

Without the application of ultrasonic power to the reaction solution, the conversion of DCDPM is low. The reaction follows a pseudo-first-order rate law, and the conversion is increased with higher ultrasonic frequency, indicating that ultrasonic waves enhance the nucleophilic substitution. The chemical effects of the ultrasound can be attributed to intense local conditions generated by cavitational bubble dynamics, i.e., the nucleation, formation, disappearance, and coalescence of vapor or gas bubbles in the ultrasonic field. However, in the phase-transfer catalytic reaction, rate enhancements are typically due to mechanical effects, mainly through an enhancement of mass transfer. The use of sonication techniques for chemical synthesis has also attracted considerable interest in recent years, because they can enhance the selectivity and reactivity, increase the chemical yields, and shorten the reaction time. In addition, there is no decomposition of phase-transfer catalysts under the experimental conditions. In this work, we found that the $k_{app,1}$ values with ultrasonic conditions (unconventional method) under the present experimental conditions are higher than those of the silent conditions (conventional method) (Wang & Rajendran, 2006; Wang &

Rajendran, 2007b, 2007c; Wang and Chen, 2010; Vivekanand & Wang, 2011). Similar increase in rate constant values was observed on varying ultrasonic power (Fig. 8). The corresponding $k_{app,1}$ values are shown in Table 2.

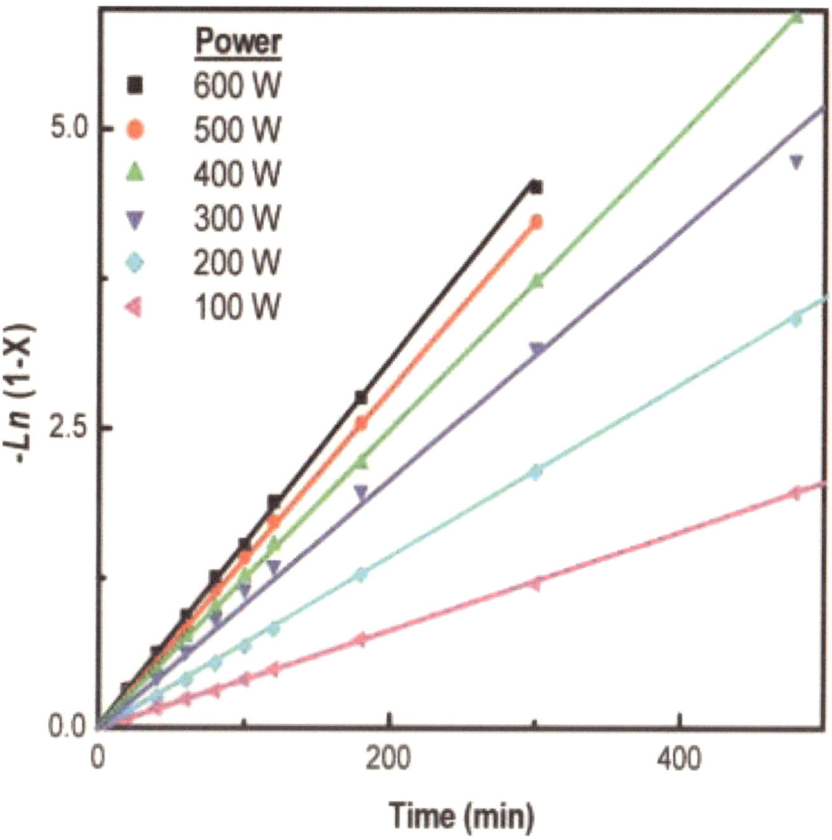

Fig. 8. Effect of the ultrasonic power on the conversion of DCDPM. Conditions: 0.228 g of TBAB, 90 mmol of methanol, 2.75 mmol of DCDPM, 40 mL of chlorobenzene, 5 g of KOH, 10 mL of water, 0.1 g of toluene, 400 rpm, 60 °C, 40 kHz. (Reprinted with permission from Wang & Chen, 2009. Copyright (2011)

Ultrasonic Power (W)	100	200	300	400	500	600
k_{app} ($\times 10^3$, min^{-1})	4.0	7.1	10.3	12.4	14.2	15.2

a- Gas chromatographic analysis

Table 2. Effect of ultrasonic power on the apparent rate constants ($k_{app,1}$) for the phase-transfer catalytic reaction of MeOH and DCDPM[a]

2.3 C-alkylation

Recently, the kinetics of monoalkylation of benzyl cyanide with *n*-bromopropane (BP) has been studied under phase transfer catalysis (PTC) conditions using aqueous potassium hydroxide as the base and tetrabutylammonium bromide as phase transfer reagent under ultrasonic condition (Vivekanand & Wang, 2011) (Scheme 5). Gas chromatography analyzed reaction was carried out at 50 ^{0}C under pseudo-first-order conditions by employing *n*-bromopropane as a limiting reactant and benzyl cyanide as a excess agent (Fig. 9).

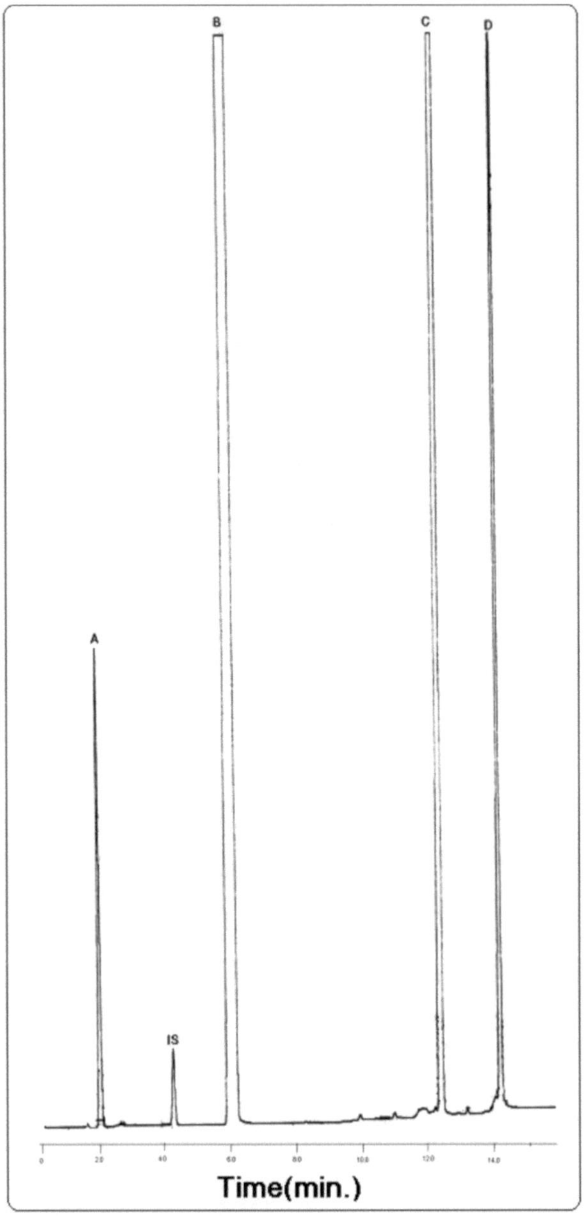

Fig. 9. GC Chromatogram of C-alkylation of benzyl cyanide. A:1-bromopropane (Retention time = 2.15 min.); IS:Toluene (Retention time = 4.40 min.); B:Chlorobenzene (Retention time = 6.19 min.); C:Benzylcyanide (Retention time = 12.52 min.); Product: 2-Phenylvaleronitrile (Retention time = 14.35 min.).

$$\underset{\mathbf{7}}{PhCH_2CN} + \underset{\mathbf{8}}{C_3H_7Br} \xrightarrow[50\ ^{\circ}C/\text{Ultrasound}]{KOH\ /\ PTC} \underset{\mathbf{9}}{PhCH(C_3H_7)CN}$$

Scheme 5. Alkylation of benzyl cyanide under PTC assisted ultrasonic condition.

The kinetic results indicates a material balance between reactant and products, i.e., the consumption of the amount of reactant (BP) equals to the sum of the generation of the amount of the product (2-phenylvaleronitrile) under ultrasonic conditions. Thus, rate of the decrease of *n*-bromopropane is consistent with the rate of production of 2-phenylvaleronitrile. Pseudo-first order kinetics was indicated by the linearity of the plot of -ln(1-X) versus time. The effect of various experimental parameters on the rate of the reaction has been studied; based on the experimental results, an interfacial mechanism was proposed. Similar PTC assisted C-alkylation reactions were analyzed by GC under various reaction conditions and an interfacial mechanism was proposed for these alkylation reactions(Vivekanand and Balakrishnan, 2009b, 2009c).

2.3.1 Effect of organic solvents

In this work, cyclohexanone, chlorobenzene, anisole, benzene, cyclohexane, were chosen as organic solvents to investigate their reactivities. From the plot of -ln(1−X) versus time, the rate constants are obtained. As shown in Table 3, the dielectric constants(ε) for these organic solvents are in the order Cyclohexanone (ε = 8.2) > Chlorobenzene (ε = 5.6) > Anisole (ε = 4.3) > Benzene (ε = 2.28) > Cyclohexane (ε = 2.02). The order of the reactivity of the reactions in these six organic solvents is Cyclohexanone > Chlorobenzene > Anisole > Benzene > Cyclohexane. The dielectric constants are usually used as the main index in choosing an appropriate organic solvent in a PTC system; i.e., the reaction rate increases with increasing dielectric constant of the organic solvent. As the dielectric constant values of solvents increases, the activity of the nucleophilic reagent and also the distance between the bromide atom and the propyl group is increased. Therefore, the rate of the reaction increases (Wang and Chen, 2009 & Wang et al., 2009).

	Solvents				
	Cyclohexane	**Benzene**	**Anisole**	**Chlorobenzene**	**Cyclohexanone**
ε^a	2.02	2.28	4.3	5.6	8.2
k_{app} **(×10⁻³, min⁻¹)**	2.2	2.9	5.7	7.0	9.8

a-Dielectric constant.

Table 3. Effect of the organic solvents on the apparent rate constants (k_{app}) under ultrasonic condition: 21.93 × 10^{-2} mol of benzyl cyanide, 10g of KOH, 20 mL of H_2O, 0.1 g of internal standard (toluene), 20 mL of solvent, 21.96 × 10^{-3} mol of *n*-bromolpropane, 0.3 g of TBAB, 600 rpm, 50 °C; under ultrasound conditions (50 kHz, 300 W). (Reprinted with permission from Vivekanand & Wang, 2011. Copyright (2011) Elsevier).

2.4 Suzuki cross coupling reaction

Suzuki cross coupling reaction (Miyaura and Suzuki, 1995; Corbet and Mignani, 2006) has been recognized as a powerful and convenient tool for the carbon–carbon bond forming methods in the synthesis of pharmaceutical agents, organic materials, as well as natural products(Tomori, et al., 2000 & Kertesz et al., 2005). As a consequence, a considerable number of homogenous palladium catalysts have been used to obtain high yields of desired product.

A good number of Suzuki cross coupling reaction have been realized in organic solvents because of the solubility of typical reaction components like aryl, allyl or benzyl halides and boronic acids as well as their coupling products. In recent times, there has been great interest in developing green chemical reactions that make carbon-carbon bonds using water as a solvent and applying these reactions in academic and industrial lab settings(Deveau & Macdonald, 2004; Li, 2005; Liu, et al., 2006). Hence researchers started do deal with the coupling reaction in aqueous two-phase systems (Genet et al., 1995) with water-soluble palladium complexes as catalysts allowing easy separation of the water phase which contains the palladium catalyst and a strong base to bind the formed hydrogen halide. In such biphasic systems, phase-transfer reagents were added to promote the transport of the water-soluble palladium catalyst to the interface of the reactant. Enhancement effect on the reaction rate was observed depending on the structure of the added amphiphiles. Thus coupling of broad range of aryl halides with organoboron compounds can be readily promoted by use of a phase transfer catalyst in a biphasic solvent system (Y.G.Wang et al., 2007 & Sahu et al., 2009). Specifically, this procedure is efficient for the cross-coupling of aryl halides with electron-withdrawing substituents and sterically demanding substituents(Miura et al., 2007). The phase-transfer catalyst system efficiently promotes the cross-coupling of electronic variation in the aryl halides.

Paetzold and Oehme (2000) reported the palladium-catalysed cross coupling of 1-iodoanisole with phenylboronic acid in an aqueous solution of sodium carbonate in the presence of different phase transfer catalysts (Scheme 6). The coupling reaction of analyzed by GLC (column HP 1; program: 2 min at 50 ^{0}C then 10^{0}C/min up to 260 ^{0}C). Gas Chromatography analysis revealed that on increasing concentration of the PTC, the rate of the reaction was increased and the formation of byproducts was suppressed.

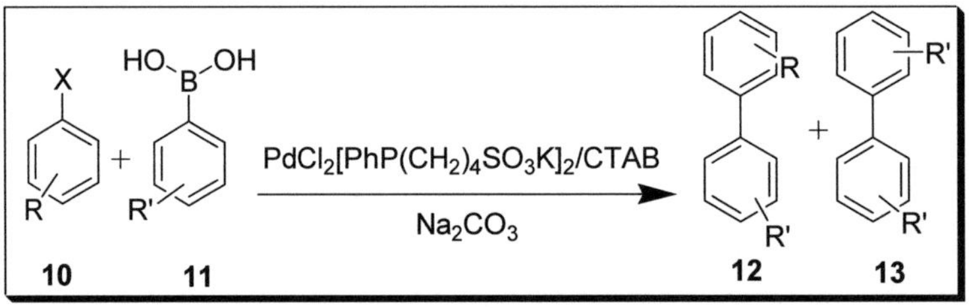

Scheme 6. Palladium-catalyzed cross coupling of iodoanisole with phenylboronic acid in the presence of PTC.

Majority of PTC's gave yields of >90% except cetylammonium tetrafluoroborate (Table 4; entry 1) which inhibits the reaction. The amphiphiles with low hydrophilic lipophilic balance (Table 4; entries 2, 5-9) and short chain amphiphiles (Table 4; entries 15 and 16) gave lower activities. Not only, the cetyltrimethylammonium bromide was favored as phase transfer reagent (Table 4; entry 10), but also zwitterionic amphiphiles, e.g., alkyldimethylammonium propane sulfonates (Table 4; entries 3 and 14) are excellent promoters. Dependence of anion was observed by authors while tetraalkylammonium salts were employed as PTC's (Table 4; entries 3, 4, 10-14). In addition to yield enhancement, the authors also reported that in presence of PTC, the selectivity increases by the suppression of formation of the biphenyl side product (**13**). Further the authors explored the reaction in presence of supported detergents(Paetzold et al., 2004).

Entry No.	PTC	Yield (%)	
		60 Min.	360 min.
1	Cetylammonium tetrafluroborate	20	47
2	Polyoxyethylene(16)methylester C_{12}	74	95
3	3-Dodecyltrimethylammonium propane sulfonate	89	94
4	Dodecyltrimethylammonium bromide	84	94
5	N-Lauroyl-alanine, sodium salt	85	89
6	N-Lauroyl-proline, sodium salt	84	89
7	Polyoxyethylene(20)sorbitanmonolaurate	78	26
8	Polyoxyethylene(6)lauryl ether	67	75
9	Polyoxyethylene(10)lauryl ether	70	78
10	Cetyltrimethylammonium bromide	98	99
11	Cetyldimethylammonium bromide	89	94
12	Cetyltrimethylammonium tetrafluoroborate	77	92
13	Cetyltrimethylammonium sulfate	78	91
14	3-Cetyldimethylammonium propane sulfonate	90	98
15	Tetra-n-butylammonium bromide	73	88
16	Benzyltrimethylammonium bromide	70	86
17	No Catalyst	65	85

Table 4. Effect of a phase-transfer catalyst in a biphasic solvent system on the Suzuki cross-coupling reactions

The coupling reaction of iodobenzene and phenylboronic acid occurs in aqueous medium in presence of palladium complexes and calix[*n*]arenes with good yields(Baur et al., 2001). The influence of water soluble macrocycles (Fig.10; 14-19) on the Suzuki reaction was analyzed by gas chromatography. The kinetic runs of aryl-aryl coupling reactions were stopped at small levels of conversion to exclude product inhibition effects and the yields were determined by GC. For comparison purpose, control experiments were carried out under identical conditions without addition of any macrocycle (Table 5, entry 9). The initial yields were found to be 4 to 8 times higher compared to the uncatalyzed reaction. Authors attributed the observed increase of initial yields to balance of binding ability of the host molecule and phase transfer properties. In order to avoid competition of organic solvent with the starting materials, apart from iodobenzene no additional organic solvent was employed in the investigation. Of the two bases compared, Cs_2CO_3 was found to be superior to diisopropylamine. All the calixarenes were found to be more superior to β-cyclodextrin. The mono-substituted calyx[4]arene **15**, the macrocyclic system with lowest solubility, was found to be more active than other tested macrocyclic systems.

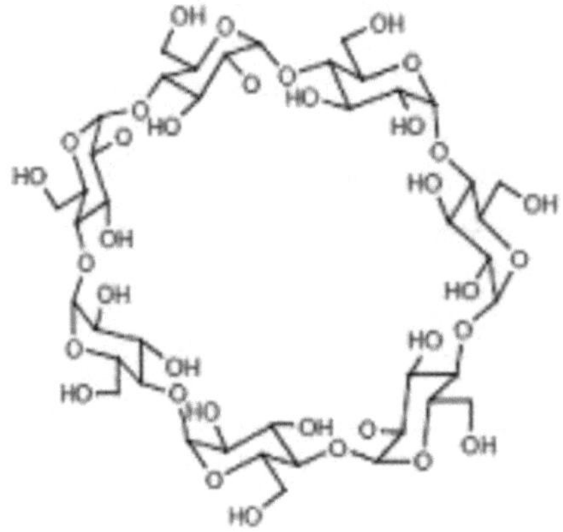

14

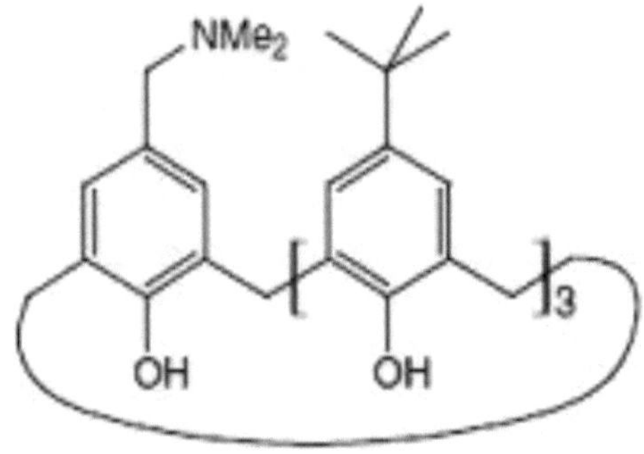

15

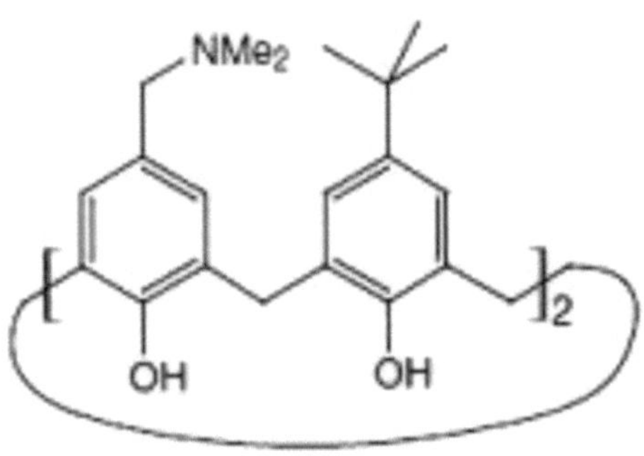

16

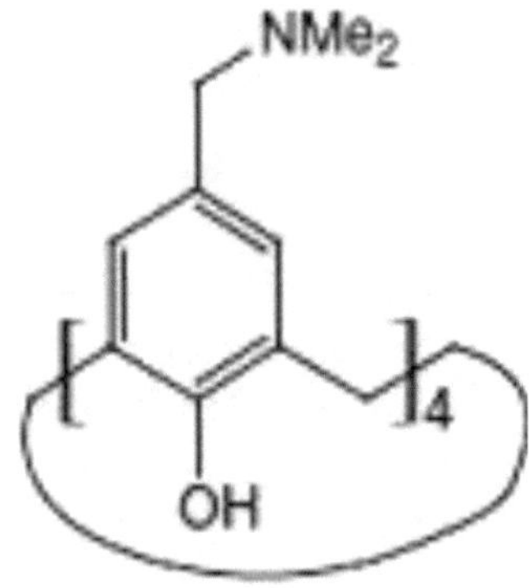

17

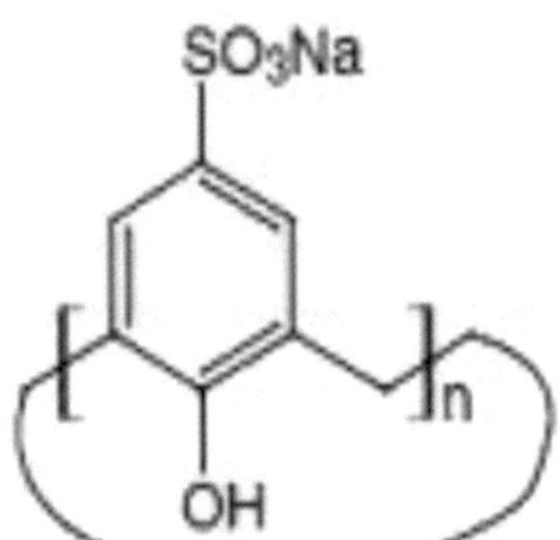

18(n=4)

19(n=6)

Fig. 10. Structure of various calixarenes.

Entry No.	Macrocycle (equiv.)	Yield (%)		
		$HNPr_2$	Cs_2Co_3	No Base
1	18 (10 mol %)	2.4	-	-
2	19 (10 mol %)	2.9	-	-
3	19 (25 mol %)	2.6	-	-
4	17 (10 mol %)	5.8	8.1	1.1
5	17 (25 mol %)	7.5		
6	16 (10 mol %)	2.2	5.5	0.1
7	15 (10 mol %)	3.6	12.7	0.1
8	14 (10 mol %)	2.1	3.4	-
9	-	≡1.0	3.2	-

Table 5. Suzuki coupling reaction in presence of different macrocycles.

2.5 Epoxidation

The high-additive-value epoxides are extensively used in insulating materials, adhesives, coating materials, construction materials and electronic parts. Due to growing attraction of PTC assisted epoxidation reactions, we examined the epoxidation of dicyclopentadiene (**19**) in presence of sodium tungstate, phosphoric acid and hydrogen peroxide under phase transfer catalysis conditions by gas chromatography(Wang et al., 2004). High conversion of dicyclopentadiene and a trace amount of by-product were obtained.

The kinetics of the reaction was carried out in a 150 ml three-necked Pyrex flask, which permitted agitating the solution, inserting thermometer, taking samples and adding the feed. Known quantities of sodium tungstate and phosphoric acid were completely dissolved in hydrogen peroxide aqueous solution for preparing the active catalyst. The solution was put into the reactor, which was submerged into a well-controlled temperature water bath within ± 0.1 ◦C. Then, measured quantities of Aliquat 336 in chloroform, dicyclopentadiene in chloroform and biphenyl (internal standard) were introduced to the reactor to start the reaction. Samples were collected from the organic layer of the mixture (by stopping the stirring for 10–15 s each time) at regular time intervals. The samples were analyzed by gas chromatography (Shimadzu, 17A) for the three products; 1,2-epoxy-3a,4,7,7a-tetrahydro-4,7-methano-indene (**20**), 5,6-epoxy-3a,4,5,6,7,7a-hexahydro-4,7-methano-indene (**21**) and 1,2,5,6-diepoxy-octahydro-4,7-methano-indene (**22**). Gas chromatography (GC) mass analysis reveals molecular ion peaks at 148 (M^+) (Fig. 11a) , 148(M^+) (Fig. 11b) and 164 (M^+) (Fig. 11c) for compounds **20, 21** and **22** respectively. The GC analyzing conditions were:

- Column: 30m x 0.525mm i.d. capillary column containing 100% poly(dimethylsiloxane)
- Injection temperature: 250^{0}C
- Carrier gas: N_2 at a flow rate of 20 mL/min
- Elution time of reactant A and products B, C, D: 3.82, 6.67, 7.24, and 9.43 min, respectively
- Detector: flame ionization detector (FID)

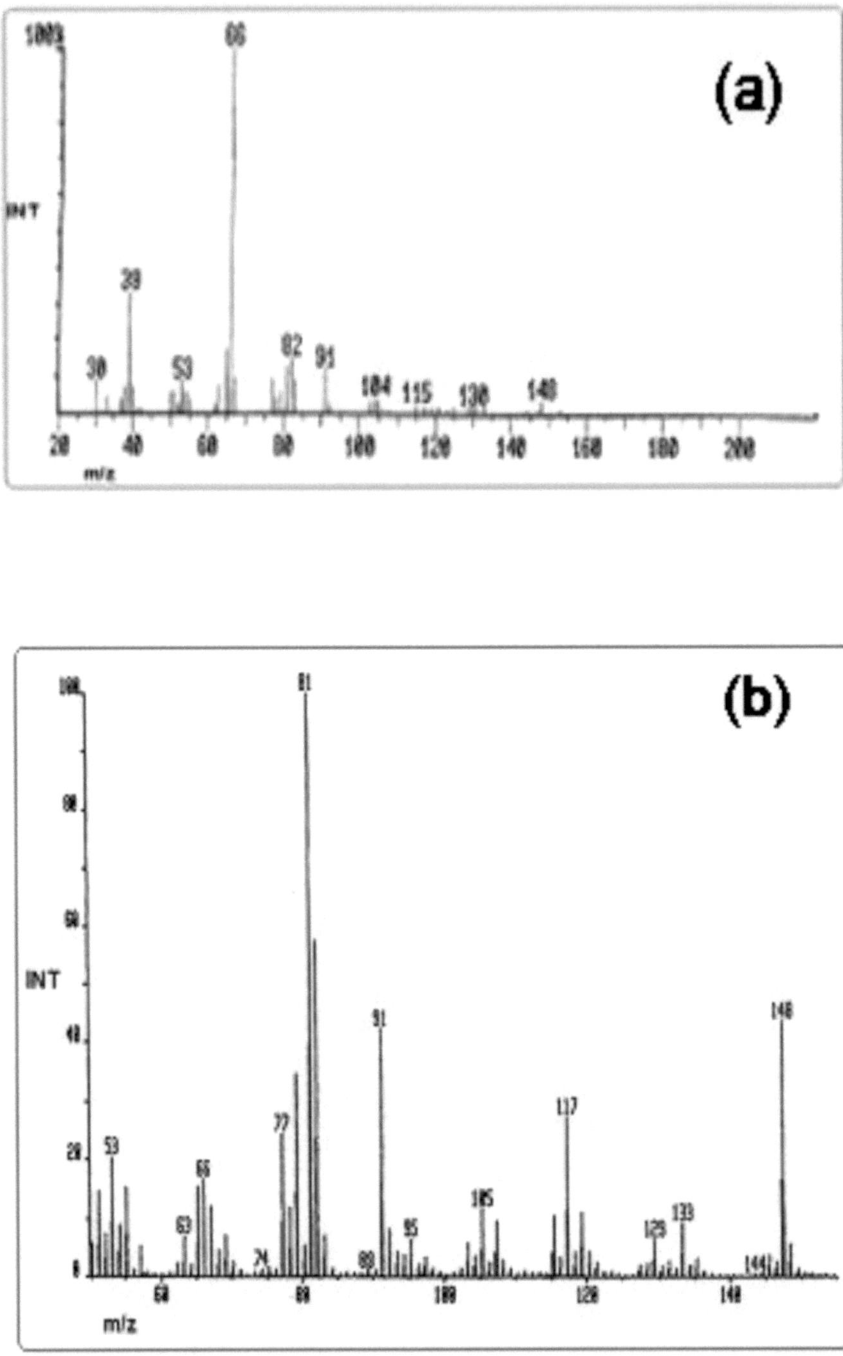

Fig. 11. Continued

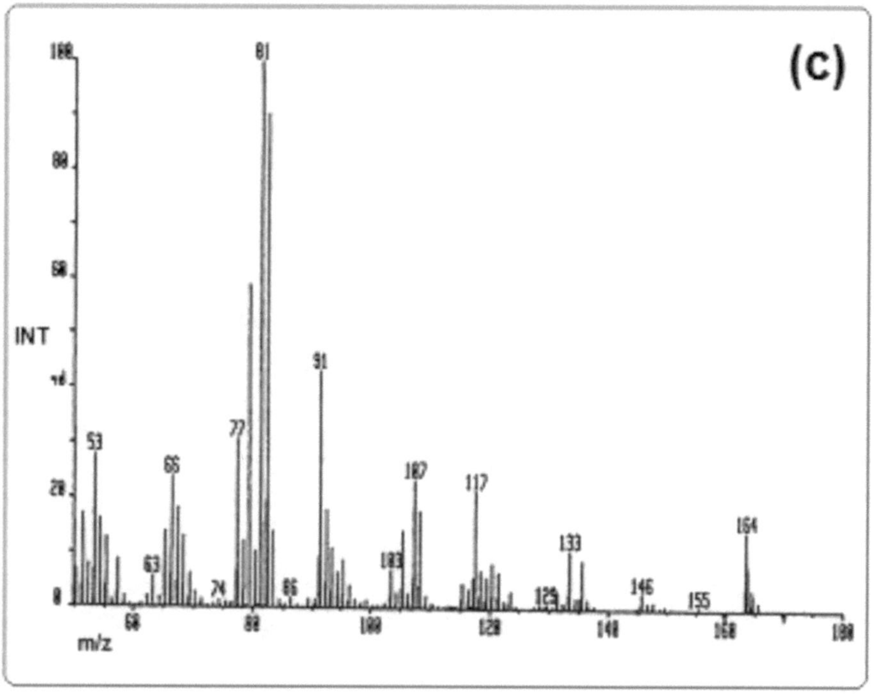

Fig. 11. GC-MS analysis of (a) 1,2-epoxy-3a,4,7,7a-tetrahydro-4,7-methano-indane(**20**), (b) 5,6-epoxy-3a,4,5,6,7,7a-hexahydro-4,7-methano-indene(**21**);(c) 1,2,5,6-diepoxy-octahydro-4,7-methano-indene (**22**).

Initially, from the ion exchange reaction (between hydrogen peroxide, phosphoric acid, sodium tungstate and quaternary ammonium salt in the aqueous phase), the active catalyst $Q_3\{PO_4[W(O)(O_2)_2]_4\}$ was generated. In general, there are eight active oxygen atoms on each molecule of the active catalyst $Q_3\{PO_4[W(O)(O_2)_2]_4\}$. Nevertheless, we assume that each time only one active oxygen atom of the active catalyst is consumed in reacting with dicyclopentadiene in each reaction step. The remaining seven oxygen atoms are no longer active and needed for regeneration, i.e., the molecular formula of the catalyst after reaction is assumed to be $Q_3\{PO_4[W(O)(O_2)_2]_3[W(O)(O_2)(O)]\}$. Thus, three products, which include the epoxidation of two single-site double bonds and one two-site double bond of dicyclopentadiene molecule, are produced. The reaction mechanism is thus proposed as shown in Scheme 7.

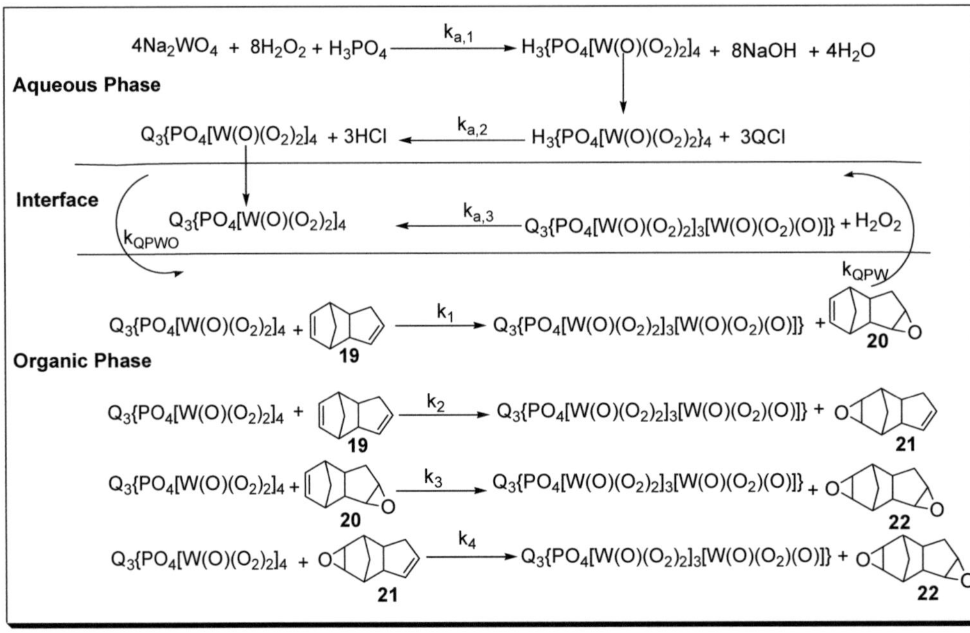

Scheme 7. Mechanism of dicyclopentadiene's epoxidation under PTC conditions. (Reprinted with permission from Wang, et al., 2004. Copyright (2011) Taylor and Francis).

where $k_{a,1}$, $k_{a,2}$, and $k_{a,3}$ are the three aqueous-phase intrinsic rate constants, k_1, k_2, k_3, and k_4 are the four organic-phase intrinsic rate constants and k_{QPW} and k_{QPWO} are the mass transfer coefficents of the regenerated catalyst $Q_3\{PO_4[W(O)(O_2)_2]_3[W(O)(O_2)(O)]\}$ and the active catalyst $Q_3\{PO_4[W(O)(O_2)_2]_4\}$, respectively. Thus, the entire reaction involves the following steps:

- the ion exchange, the complex reaction in the aqueous phase,
- formation of the active catalyst at the interface between two phases,
- the epoxidation in the organic phase, and
- the mass transfer of active catalyst.

Thus based on the experimental result, the conversion follows pseudo first-order rate law, i.e.,

$$\frac{d[DCPDI]_{org}}{dt} = k_{app}[DCPDI]_{org} \quad (25)$$

where $[DCPDI]_{org}$ is the concentration of dicyclopentadiene in the organic phase. The subscript "*org*" denotes the characteristics of the species in the organic phase. In the beginning of experiment, we have

$$t = 0, [DCPDI]_{\mathrm{org}} = [DCPDI]_{\mathrm{org,\, i}} \quad (26)$$

The subscript "*i*" denotes the characteristics of the species at the initial condition. On solving equations (25) and (26), we have

$$-\ln(1-X) = k_{app}t \tag{27}$$

where k_{app} is the apparent rate constant. X is the conversion of dicyclopentadiene, i.e.,

$$X = 1 - \frac{[DCPDI]_{org}}{[DCPDI]_{org,i}} \tag{28}$$

Assuming that the individual organic-phase reaction follows pseudo first-order rate law, we have

$$-\frac{d[DCPDI]_{org}}{dt} = -(k_1 + k_2)[QPWO]_{org}[DCPDI]_{org} \tag{29}$$

where $[QPWO]_{org}$ is the concentration of the active catalyst in the organic phase QPWO (i.e., $Q_3\{PO_4[W(O)(O_2)_2]_4\}$).

For a constant concentration of active catalyst in the organic phase, Eq (28) is reduced to

$$k_{app} = (k_1 + k_2)[QPWO]_{org} \tag{30}$$

As shown in the above equation, k_{app} represents the sum of two apparent rate constants of the two primary reactions, which are indicated in the reaction mechanism. From Eq (27), the apparent rate constant k_{app} can be obtained by a plot of - ln(1-X) versus time.

2.5.1 Effect of the amount of Aliquat 336

As shown in Table 6, the apparent rate constant value increases sharply with the increase in the amount of Aliquat 336 only upto 0.95×10^3 mol. Nevertheless, the apparent rate constant does not continue to increase when the catalyst loading exceeded 0.95×10^3 mol. We attribute the aforesaid fact to limited quantity of the active catalyst that is generated using a limiting amount of Na_2WO_4 and H_3PO_4.

Generally, only the active catalyst $Q_3\{PO_4[W(O)(O_2)_2]_4\}$ promotes the reaction. The production of the active catalyst from the reaction of quaternary ammonium salt, sodium tungstate, hydrogen peroxide, and phosphoric acid takes place only in a stoichiometric quantity. The free sodium tungstate, phosphoric acid, quaternary ammonium salt and hydrogen peroxide do not enhance the reaction. Consequently, increasing the amount of Aliquat 336 does not enhance the reaction. We observed, the products **(20)**, **(21)** and **(22)** being produced when an appropriate amount of Aliquat 336 was employed. On the other hand if a small amount of Aliquat 336 was used, only products **(20)** and **(21)** were produced.

2.5.2 Effect of the H_2O_2

Table 7 indicates the effect of amount of H_2O_2 on the rate of the reaction. The rate decreased with the increase in the volume of H_2O_2. Nonetheless, this variation is not significant. The foremost cause is probably that the oxidation of dicyclopentadiene by free H_2O_2 takes place when a larger volume of H_2O_2 is employed. The active catalyst distributes between aqueous and organic phases, and most of the active catalyst is soluble in the organic phase. Therefore, the active catalyst of a slight portion dissolves in the aqueous phase by increasing the volume of H_2O_2. Hence, the concentration of the active catalyst in the organic phase is slightly decreased by increasing the volume of hydrogen peroxide. The reaction between the active catalyst and the reactant takes place in the organic phase. Thus, the reaction rate slightly decreased with the increase in the volume of hydrogen peroxide due to low concentration of active catalyst in the organic phase(Wang and Rajendran, 2007a,b & Wang and Chen, 2008).

Amount of Aliquat 336 x 10^3 (mol)	1.24	1.07	0.95	0.77	0.58	0.39	0.21	0.08	0
$k_{1,app}$ **x 10^2(min^{-1})**	2.79	2.99	3.15	3.15	1.61	1.91	1.17	0.44	0.12
$k_{2,app}$ **x 10^2(min^{-1})**	1.51	1.57	1.70	1.70	0.81	0.96	0.56	0.19	0
$k_{3,app}$ **x 10^2(min^{-1})**	0.63	0.69	0.79	0.79	0.20	0.42	0.25	0	0
$k_{4,app}$ **x 10^2(min^{-1})**	1.11	1.24	1.41	1.41	0.55	0.82	0.40	0	0

Table 6. Influence of the amount of catalyst on the conversion of dicyclopentadiene based on GC analysis

Volume of H_2O_2(mL)	65	55	45	35	25
$k_{1,app}$ **x 10^2(min^{-1})**	1.97	2.13	2.26	2.71	2.37
$k_{2,app}$ **x 10^2(min^{-1})**	1.23	1.35	1.39	1.61	1.53
$k_{3,app}$ **x 10^2(min^{-1})**	0.50	0.57	0.63	0.67	0.62
$k_{4,app}$ **x 10^2(min^{-1})**	0.90	1.04	1.04	1.22	1.11

a-by GC analysis

Table 7. Influence of the volume of hydrogen peroxide on the conversion of dicyclopentadiene[a]

Previously, we reported (Wang & Huang, 2003) GC analyzed comparison of epoxidation conversion of olefins under phase transfer catalysis conditions. The results are shown in Table 8. In Olefins (**23-38**), the double bond in the ring is more easily oxidized than that of the terminal one using hydrogen peroxide as an oxidant. We compared the epoxidation results of olefins (**27**), (**29**) & (**31**) with those of olefins (**34**), (**36**) & (**38**). The results indicates higher conversion of olefins with double bond in five-carbon, six-carbon and eight-carbon rings than that of the aliphatic terminal bond. With minimum steric hindrance, only two hydrogen atoms appear on the double bond of the ring. On the other hand, there is larger steric hindrance for the double bond of the chain end when using the large-sized catalyst ($Q_3\{PO_4[W(O)O_4]_4\}$). Due to this phenomenon, it is difficult for the oxygen atom in the active complex catalyst to combine with the terminal carbon-carbon double bond.

As stated previously, the double bond of the ring can easily be epoxidized. Therefore, conversion of dicyclopentadiene (**19**) is larger than that of the olefins (**23**), & (**25**), which have only one double bond in the ring. Three different products were obtained from olefin (**19**) for the two double bonds in the ring. Since, six-carbon ring epoxide is more stable, conversion of olefin (**27**) is higher than olefin (**29**). On comparing the conversion of olefins (**27**), (**29**) & (**30**), we found that the conversion of 1,5- cyclooctadiene (100%) was the highest and the conversion of cyclohexene (92%) was the lowest. The reason for this is that the six-carbon ring is the most stable, whereas the eight-carbon ring is the most unstable.

The conversion of straight chain compounds [(**34**), (**36**) & (**38**)] was much lower than those of the above reactants [(**19**), (**23**) (**25**), (**27**), (**29**), & (**31**)]. Comparing the results of (**34**) and (**36**), the conversion of 1-hexene was larger than that of 1-octene. Generally, the double bonds of the six-carbon chain have more chances to be attacked by the active catalyst than those of the 8-carbon chain. In the case of (**38**), there are two double bonds on the chain end of 1,7-octadiene. However, epoxidation of 1,7-octadiene gave only one product because of the short reaction time. Nevertheless *bis*-epoxy product was noticed on extending the reaction time.

Entry	Olefin	Product	Conversion[a]
1	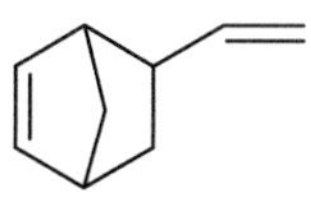 **23**	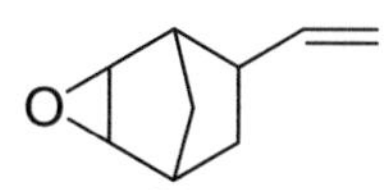 **24**	75
2	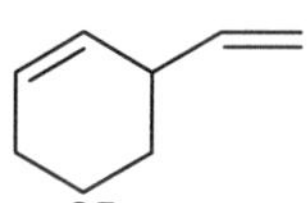 **25**	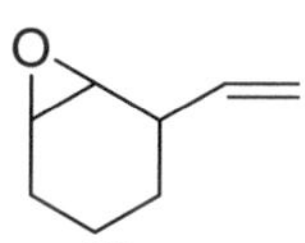 **26**	92
3	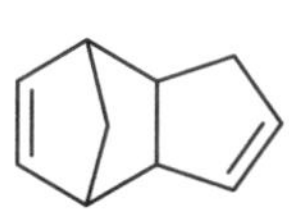 **19**	 **20**  **21** 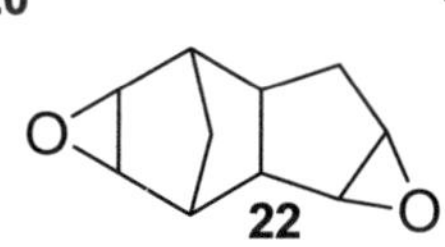 **22**	98
4	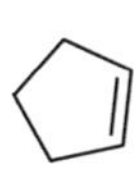 **27**	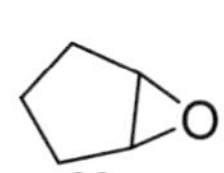 **28**	98
5	 **29**	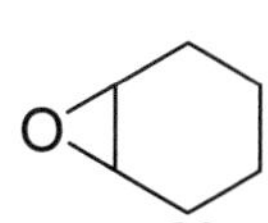 **30**	92
6	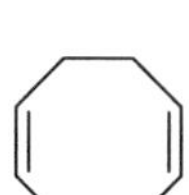 **31**	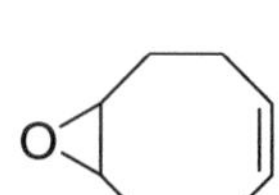 **32** 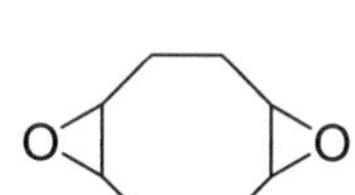 **33**	100

7	 34	 35	25
8	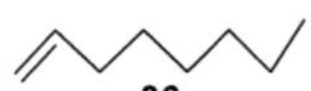 36	 37	10
9	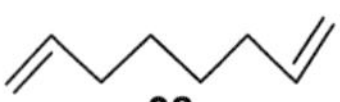 38	 39	20

a Gas chromatography (Shimadzu GC 9A with FID using 100% poly(dimethylsiloxane), capillary column, 30m_0.525 mm and nitrogen as the carrier gas).

Table 8. Structural influence of different olefins on their epoxidation under PTC conditions (Wang & Huang, 2003) analyzed by GC [a]. (Reprinted with permission from Wang & Huang, 2003. Copyright (2011) Springer).

3. Conclusions

Current chapter describes recent developments and a perspective in kinetics of phase transfer catalysis assisted organic reactions analyzed by Gas Chromatography. The combination of PTC with sonochemistry can also be an interesting approach for determining efficient reaction conditions, provided that PTC remains stable under exposure to ultrasounds. Research reports from leading laboratories that touch on all the themes noted herein and many others are assembled. The topics are organized by five broad groupings, *viz.*, *N*-alkyaltion, *O*-Alkylation, *C*-alkylation, Suzuki coupling and epoxidation. The N-alkylation study reveals that the stirring speed has positive influence on rate of the reaction only up to 400 rpm and the favorable quaternary ammonium salt for the reaction will be the one with larger carbon numbers. In the ultrasonic irradiation assisted synthesis of dimethoxydiphenylmethane, higher the ultrasonic frequenzy and power, higher will be the rate constant values. C-alkylation study indicates the influence of dielectric constant values on the rate of the reaction. GC-analyzed Suzuki coupling reactions indicates the application of PTC in these reactions. Kinetics of epoxidation of dicyclopentadiene shows that the rate of the reaction depends only upto certain amount of catalyst. Nevertheless the volume of hydrogen peroxide has a negative influence on the rate of the reaction. Thus, PTC approach leads in establishing genuinely sustainable chemical industrial processes within the context of the forthcoming paradigm shift in worldwide production of highly valuable substances.

4. Acknowledgments

We gratefully acknowledge support of this work by the National Science Council, Taiwan (NSC), under several grants.

5. References

Awasthi, S. Rao A. N.,& Ganesan, K.. (2009). An environmental-benign approach for the synthesis of alkylthiocyanates. *J. Sulfur Chem.*, Vol.30, No.5, pp.513–517, ISSN: 1741-5993

Baelen, G.V., Maes, B. U. W. (2008). Study of the microwave-assisted hydrolysis of nitriles and esters and the implementation of this system in rapid microwave-assisted Pd-catalyzed amination. *Tetrahedron, Vol.* 64, No.23, pp.5604-5619, ISSN: 0040-4020

Barbasiewicz, M., Marciniak, K., & Fedoryn´ski, M. (2006). Phase transfer alkylation of arylacetonitriles revisited.*Tetrahedron Lett.* Vol.47, No.23, pp.3871-3874, ISSN: 0040-4039

Baur, M., Frank, M., Schatz, J., & Schildbach, F. (2001).Water-soluble calyx[*n*]arenes as receptor molecules for non-polar substrates and inverse phase transfer catalysts. *Tetrahedron,* Vol.57, No.32, pp.6985-6991, ISSN: 0040-4020

Bogdal, D., Lukasiewicz, M., Pielichowski, J., & Bednarz, S. (2005). Microwave-assisted epoxidation of simple alkenes in the presence of hydrogen peroxide. *Synth. Commun.*,Vol.35, No.23, pp.2973-2983, ISSN: 0039-7911

Bravo, R., Caltabiano, L. M., Fernandez, C., Smith, K. D., Gallegos, M., Whitehead Jr., R. D., Weerasekera, G., Restrepo, P., Bishop, A. M., Perez, J. J. , & Needham, L.L., Barr, D. B. (2005). Quantification of phenolic metabolites of environmental chemicals in human urine using gas chromatography–tandem mass spectrometry and isotope dilution quantification. *J CHROMATOGR B.*, Vol.820, No.2, pp.229-236, ISSN: 1570-0232

Cardador, M. J., Serrano, A., & Gallego, M. (2008). Simultaneous liquid–liquid microextraction/methylation for the determination of haloacetic acids in drinking waters by headspace gas chromatography. *J CHROMATOGR A.*, Vol.1209, No.1-2, pp.61–69, ISSN: 0021-9673

Chatti, S., Bortolussi, M., Loupy, A., Blais, J. C., Bogdal, D., & Majdoub, M. (2002). Efficient synthesis of polyethers from isosorbide by microwave-assisted phase transfer catalysis. *Eur. Poly. J.* Vol.38, No.9, pp.1851-1861, ISSN: 0014-3057

Chatti, S., Bortolussi, M., Bogdal, Blais, D. J. C., & Loupy, A. (2004). Microwave-assisted polycondensation of aliphatic diols of isosorbide with aliphatic disulphonylesters via phase-transfer catalysis. *Eur. Poly.J.* Vol.40, No.3, pp.561-577, ISSN: 0014-3057

Corbet, J. P., Mignani, G. (2006). Selected patented cross-coupling reaction technologies. *Chem. Rev.*, Vol.106, No.7, pp.2651-2710, ISSN: 0009-2665

Deveau, A. M., & Macdonald, T. L. (2004). Practical synthesis of biaryl colchicinoids containing 3′,4′-catechol ether-based A-rings via Suzuki cross-coupling with ligandless palladium in water. *Tetrahedron Lett.*, Vol.45, No.4, pp.803-807, ISSN: 0040-4039

Devulapelli, V. G., & Weng, H.S. (2009). Synthesis of cinnamyl acetate by solid–liquid phase transfer catalysis: Kinetic study with a batch reactor *Catal. Commun.* Vol.10, No.13 , pp.1638-1642, ISSN: 1566-7367

Fiamegos, Y. C., Kefala, A.P., & Stalikas, C. D. (2008). Ion-pair single-drop microextraction versus phase-transfer catalytic extraction for the gas chromatographic determination of phenols as tosylated derivatives. *J CHROMATOGR A.*, Vol.1190, No.1-2, pp.44-51, ISSN: 0021-9673

Fiamegos, C., Karatapanis, A., & Stalikas, C. D., (2010). Microwave-assisted phase-transfer catalysis for the rapid one-pot methylation and gas chromatographic determination of phenolics *J CHROMATOGR A.*, Vol.1217, No.5, pp.614-621, ISSN: 0021-9673

Genet, J.P., Linquist, A., Blart, E., Mouries, V., Savignac, M., & Vaultier, M. (1995). Suzuki-type cross coupling reactions using palladium-water soluble catalyst. Synthesis of functionalized dienes. *Tetrahedron Lett.,* Vol.36, No.9, pp.1443 -1446, ISSN : 0040-4039

Greiner, I., Sypaseuth, F. D., Grun, A., Karsai, E., & Keglevich, G.. (2009). The role of phase transfer catalysis in the microwave-assisted N-benzylation of amides, imides and N-heterocycles. *Lett. Org.Chem.* Vol.6, No.7, pp.529-534, ISSN: 1570-1786

Gumaste, V. K., Khan, A. J., Bhawal, B. M., & Deshmukh, A.R. A. S. (2004). Microwave assisted phase transfer catalysis: An efficient solvent free method for the synthesis of cyclopropane derivatives. *Ind. J. Chem., Sec. B.,* Vol.43B, No.2, pp.420-423, ISSN: 0376-4699

Hejchman, E., Maciejewska, D., & Wolska, I. (2008). Microwave-assisted synthesis of derivatives of khellinone under phase-transfer catalytic conditions *Monatshefte fuer Chemie.,* Vol.139, No.11, pp.1337–1348, ISSN: 0026-9247

Iijima, T., Suzuki, N., Fukuda, W., & Tomoi, M. (1995).Toughening of aromatic diamine-cured epoxy resins by modification with N-phenylmaleimide-styrene-*p*-hydroxystyrene terpolymers. *Eur. Polym. J.,* Vol.31, No.8, pp.775-783, ISSN: 0014-3057

Jankovic, V., Mijin, D.Z., & Petrovic, S.D. (2002). Alkylation of N-substituted 2-phenylacetamides: Benzylation of N-(4-nitrophenyl)-2-phenylacetamide. *J.Serb.Chem.Soc.,* Vol.67, No.6, pp.373-379, ISSN: 0352-5139

Kertesz, M., Choi, C. H., & Yang, S. (2005). Conjugated polymers and aromaticity. *Chem. Rev.,* Vol.105, No.10, pp.3448-3481, ISSN: 0009-2665

Lancaster, M. (2002). Green Chemistry: An Introductory Text- Royal Society of Chemistry, pp 119-122. ISSN: 0-85404-6208

Langmuir, M. E., Yang, J. R., Moussa, A. M., Laura, R., & Lecompte, K. A. (1995). New naphthopyranone based fluorescent thiol probes. *Tetrahedron Lett.,* Vol.36, No.23, pp.3989 -3992, ISSN: 0040-4039

Li, C.J. (2005). Organic reactions in aqueous media with a focus on carbon-carbon bond formations: A decade update. *Chem. Rev.,* Vol.105, No.8, pp.3095-3166, ISSN: 0009-2665

Lin, F. M., Wu, H. S., Kou, H. S., & Lin, S. J. (2003). Highly sensitive analysis of iodide anion in seaweed as pentafluorophenoxyethyl derivative by capillary gas chromatography. *J. Agric. Food Chem.* Vol.51, No.4, pp.867-870, ISSN: 0021-8561

Liu, L., Zhang, Y.,& Xin, B. (2006). Synthesis of biaryls and polyaryls by ligand-free suzuki reaction in aqueous phase. *J. Org. Chem.,* Vol.71, No.10, pp.3994-3997, ISSN: 0022-3263

Luo, C. Lü, J., Cai, C., & Qü. W. (2004). A polymer onium acting as phase-transfer catalyst in halogen-exchange fluorination promoted by microwave. *J. Fluor.Chem.,* Vol.125, No.5, pp.701-704. ISSN: 0022-1139

Mahdavi. H., & Tamami, B.(2005). Synthesis of 2-nitroalcohols from epoxides using quaternized amino functionalized cross-linked polyacrylamide as a new polymeric phase transfer catalyst. *REACT FUNCT POLYM,* Vol. 64, No.3, pp.179-185,. ISSN: 1381-5148

Makosza, M. (2000). Phase-transfer catalysis. A general green methodology in organic synthesis. *Pure Appl. Chem.,* Vol.72, No.7, pp.1399-1403, ISSN: 0033-4545

Masuno, M.N., Young, D.M., Hoepker, A.C., Skepper, C.K., & Molinski, T.F. (2005). Addition of Cl_2C: to (−)-*O*-Menthyl Acrylate under Sonication−Phase-Transfer

Catalysis. Efficient Synthesis of (+)- and (−)-(2-Chlorocyclopropyl)methanol. *J. Org. Chem.*, Vol.70, No.10, pp.4162-4165, ISSN: 0022-3263

Mijin, D.Z., Jankovic, V., & Petrovic, S.D. (2004). Alkylation of N-substituted 2-phenylacetamides: Benzylation of N-(4-chlorophenyl)-2-phenylacetamide. *J.Serb.Chem.Soc.*, Vol.69, No.2, pp.85-92, ISSN: 0352-5139

Mijin, D.Z., Prascevici, M., & Petrovic, S.D. (2008). Benzylation of N-phenyl-2-phenylacetamide under microwave irradiation. *J.Serb. Chem. Soc.*, Vol.73, No.10, pp.945-950, ISSN: 0352-5139

Mikia, A., Tsuchihashia, H., Yamanob, H., & Yamashita, M. (1997). Extractive pentafluorobenzylation using a polymeric phase-transfer catalyst: a convenient one-step pretreatment for gas chromatographic analysis of anionic compounds. Anal. Chim. Acta., Vol.356, No.2-3, pp.165-175,. ISSN: 0003-2670

Miura, M., Koike, T., Ishihara, T., Sakamoto, S., Okada, M., Ohta, M., & Tsukamoto, S.-i. (2007). One-pot preparation of unsymmetrical biaryls via Suzuki Cross-Coupling reaction of aryl halide using phase-transfer catalyst in a biphasic solvent system. *Syn. Commun., Vol.*37, No.5, pp.667-674, ISSN 0039-7911

Miyaura, N., Suzuki, A. (1995). Palladium-catalyzed cross-coupling reactions of organoboron compounds. *Chem. Rev.,* Vol.95, No.7, pp.2457-2483, ISSN: 0009-2665

Ohkubo, M., Nishimura, T. , Jona, H., Honma, T., & Morishima, H. (1996). Practical synthesis of indolopyrrolocarbozoles. *Tetrahedron,* Vol.52, No.24, pp.8099-8112, ISSN: 0040-4020

Paetzold, E., & Oehme, G.. (2000). Efficient two-phase Suzuki reaction catalyzed by palladium complexes with water-soluble phosphine ligands and detergents as phase transfer reagents. *J. Mol. Catal. A: Chem. Vol.*152, No. 1-2, pp.69-76, ISSN: 1381-1169

Paetzold, E., Jovel, I., & Oehme, G.. (2004). Suzuki reactions in aqueous multi-phase systems promoted by supported detergents. *J. Mol. Catal. A: Chem. Vol.*214, No. 2, pp.241-247, ISSN: 1381-1169

Sahu, K. B., Hazra, A., Paira, P., Saha, P., Naskar. S., Paira, R. , Banerjee, S., Sahu, N. P., B.Mondal, N., Luger, P., & Weber, M. (2009). Synthesis of novel benzoxazocino quinoliniums and quinolones under PTC conditions and their application in Suzuki cross coupling reaction for the construction of polynuclear heteroaromatics. *Tetrahedron*, Vol.65, No.34, pp.6941–6949, ISSN: 0040-4020

Sakayanagi, M., Yamada, Y., Sakabe, C., Watanabe, K., & Harigaya, Y., (2006). Identification of inorganic anions by gas chromatography/mass spectrometry. *Forensic Sci. Int.*, Vol.157, No.2-3, pp.134-143, ISSN: 0379-0738

Sasson, Y., & Neumann, R.(1997). *Handbook of Phase Transfer Catalysis*, Blackie Academic and Professional (Chapman & Hall), ISBN: 0751402583, London.

Settimo, A.Da., Primofiore, G., Settimo, F.Da., Simorini, F. , Motta, C.La., Martinelli, A., & Boldrine, E. (1996). Synthesis of pyrrolo[3,4-*c*]pyridine derivatives possessing an acid group and their *in vitro* and *in vivo* evaluation as aldose reductase inhibitors. *Eur. J. Med. Chem.*, Vol.31, No.1, pp.49-58, ISSN: 0223-5234

Sharma, P., Kumar, A., & Sharma, M. (2006). Synthesis of 4-[2-(2-methylprop-1-enylidene)-2,3-dihydro-1H-benzimidazole-1-yl]-1-napthol *via* azo group insertion of dimethylvinylidenecarbene under phase transfers catalysis conditions. *Catal. Commun. Vol.*7, No.9, pp. 611-617. ISSN: 1566-7367.

Tomori, H., Fox, J. M., & Buchwald, S. L. (2000). An improved synthesis of functionalized biphenyl-based phosphine ligands.*J. Org. Chem.*, Vol.65, No.17, pp.5334-5341, ISSN: 0022-3263

Vivekanand, P. A., & Balakrishnan, T. (2009a). Kinetics of dichlorocyclopropanation of vinylcyclohexane catalyzed by a new multi-site phase transfer catalyst. *Cat. Commun.*, Vol.10, No. 5, pp.687-692, ISSN:1566-7367

Vivekanand, P. A., & Balakrishnan, T. (2009b). Superior catalytic efficiency of a new multisite phase transfer catalyst in the C-alkylation of dimedone - A kinetic study. *Cat. Commun.*, Vol.10, No.10, pp.1371-1375. ISSN:1566-7367

Vivekanand, P. A., & Balakrishnan, T. (2009c). Synthesis and characterization of a novel multi-site phase transfer catalyst and a kinetic study of the intramolecular cyclopentanation of indene. *Appl. Catal. A: Gen.*, Vol.364, No.1-2, pp.27-34. ISSN: 0926-860X.

Vivekanand, P. A., & Balakrishnan, T. (2009d). Catalytic potential of a new polymeranchored multisite phase tansfer catalyst in the dichlorocarbene addition to Indene. *Catal. Lett.*, Vol.13, No.3-4, pp.587-596, ISSN: 1011-372X

Vivekanand, P. A., & Balakrishnan, T. (2009e). Evaluation of catalytic efficiency of a triple-site phase transfer catalyst and the kinetics of dichlorocarbene addition to 5-vinyl-2-norbornene. *Cat. Commun.*, Vol.10, No.15, pp.1962-1966, ISSN:1566-7367

Vivekanand, P.A., & Wang, M. L. (2011). Sonocatalyzed synthesis of 2-phenylvaleronitrile under controlled reaction conditions - A kinetic study. *Ultrason. Sonochem.*, Vol.18, No.5 , pp.1241–1248, ISSN: 1350-4177

Wang, M.L., & Tseng, Y.H. (2002). Phase-transfer catalytic reaction of dibromo-*o*-xylene and 1-butanol in two-phase solution. *J. Mol. Catal. A: Chem.*, Vol.179, No.1-2, pp.17-26, ISSN: 1381-1169

Wang, M.L., Hsieh, Y.M., & Chang, R.Y. (2003). Kinetic study of dichlorocyclopropanation of 1,7-octadiene under phase-transfer catalysis conditions at high alkaline concentration *Ind. Eng. Chem. Res*,. Vol.42, No.20, pp.4702-4707, ISSN: 0888-5885

Wang, M.L., & Huang, T.H. (2003). Phase-transfer catalytic epoxidation of oleins under liquid-liquid biphasic conditions. *React. Kinet. Catal. Lett.*, Vol.78, No.2, pp.275-280, ISSN : 0133-1736

Wang, M.L., Huang. T.H., & Wu, W. T. (2004). Kinetic study of the phase transfer catalytic epoxidation of dicyclopentadiene in a two-phase medium. *Chem. Eng. Comm., Vol.* 191, No.1, pp.27-46, ISSN: 0098-6445

Wang, M.L., Chen, W.H., & Wang, F. S. (2005). Kinetic study of synthesizing N-butylphthalimide under solid-liquid phase-transfer catalysis conditions *J. Mol. Catal. A: Chem.*, Vol.236, No.1-2, pp.65-71, ISSN: 1381-1169

Wang, M.L., & Lee, Z.F. (2006). Kinetic study of synthesizing bisphenol a dially ether in a phase transfer catalytic reaction. *Bull. Chem. Soc. Jpn.* Vol.79, No.1, pp.80–87, ISSN: 0009-2673

Wang, M. L., & Rajendran, V. (2006). A kinetic study of thioether synthesis under influence of ultrasound assisted phase-transfer catalysis conditions. *J. Mol. Catal. A: Chem.*, Vol.244, No.1-2, pp.237–243, ISSN: 1381-1169

Wang, M.L., & Lee, Z.F. (2007). Reaction of 4,4′-bis(chloromethyl)-1,1′-biphenyl and phenol in two-phase medium via phase-transfer catalysis *J. Mol. Catal. A: Chem.*, Vol.264, No.1-2, pp.119-127, ISSN: 1381-1169

Wang, M. L., & Rajendran, V. (2007a). Ultrasound assisted phase-transfer catalytic epoxidation of 1,7-octadiene - A kinetic study. *Ultrason. Sonochem.*, Vol.14, No.1, pp.46-54, ISSN:1350-4177.

Wang, M. L., & Rajendran, V. (2007b). Kinetics for dichlorocyclopropanation of 1,7-octadiene under the influence of ultrasound assisted phase-transfer catalysis conditions *J. Mol. Catal. A: Chem.*, Vol.273, No.1-2, pp.5-13, ISSN: 1381-1169

Wang, M.L., & Rajendran V. (2007c). Ethoxylation of *p*-chloronitrobenzene using phase-transfer catalysts by ultrasound irradiation – A kinetic study. *Ultrason. Sonochem.*, Vol.14, No.3, pp.368–374,. ISSN: 1350-4177

Wang, M. L.,& Chen, C. J. (2008). Kinetic Study of Synthesizing 1-(3-Phenylpropyl)-pyrrolidine-2,5-dione under Solid-Liquid Phase-Transfer Catalysis. *Org. Process Res. Dev.*, Vol.12, No.4, pp. 748–754, ISSN: 1083-6160

Wang, M. L., Chen, C. J., & Wang, F. S. (2009). Kinetic study of synthesizing 1-phenyl-3-propyl ether under liquid-liquid phase-transfer catalytic conditions. *Chem. Eng. Comm.*, Vol.196, No.5, pp.573–590, ISSN: 0098-6445

Wang, M.L., & Chen, W.H. (2009). Kinetic study of synthesizing dimethoxydiphenylmethane under phase-transfer catalysis and ultrasonic irradiation. *Ind. Eng. Chem. Res., Vol.*48, No.3, pp.1376-1383, ISSN: 0888-5885

Wang, M. L.,& Chen, C. J. (2010). Kinetic Study of Synthesizing 1-(3-Phenylpropyl)pyrrolidine-2,5-dione under Solid-Liquid Phase-Transfer Catalytic Conditions Assisted by Ultrasonic Irradiation. *Org. Process Res. Dev.* Vol.14, No.3, pp.737-745, ISSN: 1083-6160

Wang, M. L., & Prasad, G.S. (2010). Microwave assisted phase-transfer catalytic ethoxylation of *p*-chloronitrobenzene—A kinetic study. *J. Taiwan Inst. Chem. Eng.*, Vol.41, No.1, pp.81-85, ISSN: 1876-1070.

Wang, Y.G., Ueda, M., Wang, X., Han, Z., & Maruoka, K. (2007). Convenient preparation of chiral phase-transfer catalysts with conformationally fixed biphenyl core for catalytic asymmetric synthesis of α-alkyl- and α,α-dialkyl-a-amino acids: application to the short asymmetric synthesis of BIRT-377. *Tetrahedron,*Vol. 63, No.26, pp.6042-6050, ISSN: 0040-4020.

Yadav, G. D., & Bisht, P. M. (2004). Novelties of microwave assisted liquid–liquid phase transfer catalysis in enhancement of rates and selectivities in alkylation of phenols under mild conditions. *Catal. Commun.* Vol.5, No. , pp.259–263, ISSN: 1566-7367

Yadav, G. D., (2004). Insight into green phase transfer catalysis. *Top. Catal.* Vol.29, No.3-4, pp.145-161, ISSN: 1022-5528

Yadav, G. D., & Badure, O. V. (2008). Selective engineering in O-alkylation of *m*-cresol with benzyl chloride using liquid–liquid–liquid phase transfer catalysis. *J. Mol. Catal. A: Chem.* Vol.288, No.1-2, pp.33-41. ISSN: 1381-1169

Yang, H.M., & Peng, G.Y. (2010). Ultrasound-assisted third-liquid phase-transfer catalyzed esterification of sodium salicylate in a continuous two-phase-flow reactor. *Ultrason.Sonochem.*, Vol.17, No.1, pp.239–245, ISSN: 1350-4177.

Yang, H.M., & Lin, D.W. (2011). Third-liquid phase-transfer catalyzed esterification of sodium benzoate with novel dual-site phase-transfer catalyst under ultrasonic irradiation. *Catal. Commun.*, Vol.14, No.1, pp.101-106. ISSN:1566-7367

Yiannis, C. F., Nanos, C. G., Vervoort, J., & Stalikas, C. D. (2004). Analytical procedure for the in-vial derivatization—extraction of phenolic acids and flavonoids in methanolic and aqueous plant extracts followed by gas chromatography with mass-selective detection. *J CHROMATOGR A.*, Vol.1041, No.1-2, pp.11–18, ISSN: 0021-9673

6

POP and PAH in Bizerte Lagoon, Tunisia

Trabelsi Souad, Ben Ameur Walid, Derouiche Abdekader,
Cheikh Mohamed and Driss Mohamed Ridha
*Laboratory of Environmental Analytical Chemistry,
Faculty of Sciences, Bizerte, Zarzouna,
Tunisia*

1. Introduction

Gas chromatography is the most recent branch of chromatography and includes all the chromatography processes in which the substance to be analyzed occurs in the gaseous or vapor state or can be converted into such a state. Although the first records of gas chromatography go back hundreds of years (Bayer, 1959), its true history began during World War II when a large industrial chemical company instituted a crash its development (Tietz, 1970). The first published work appeared in the early 1950´s based on the successful experiments by James and Martin (1952). In the years between 1952 and 1956 the early apparatus came into sight. The success of gas chromatography is due to its simplicity of operation, high separation power and speed. Its use involves versatile applications not only to the field of chemistry but also biology, medicine, industrial research and control, environmental health and scientific studies.

In the field of environmental control the use of gas chromatography has allowed specific separations and measurements of toxic compounds on a large variety of matrices. Among these hazards, persistent organic pollutants (POPs) are carbon-based organic compounds and mixtures with toxicity and environmental persistence that include industrial products and byproducts. POPs can be transported far from their sites of release by environmental media to previously pristine locations such as the Arctic. Low POP levels might be increased by biomagnification through the transmission process in the food chain. They can be easily accumulated in the organism to levels that can potentially injure human health as well as the environment (Birnbaum, 1994; Hansen, 1998). There are 12 substances or substance groups prioritized for global action in the recently signed Stockholm Convention on Persistent Organic Pollutants, developed under the auspices of the United Nations Environment Programme (UNEP).

The 12 substances, the "dirty dozen," consist of eight kinds of pesticides, including dieldrin, aldrin, endrin, chlordane, heptachlor, DDT, toxaphene, mirex, two kinds of industrial chemicals [polychlorinated biphenyls (PCBs) and hexachlorobenzene (HCB)], and two kinds of byproducts (polychlorinated dibenzofurans and polychlorinated dibenzo-p-dioxins) (Stockholm Convention on Persistent Organic Pollutants, 2001). There are four characteristic parameters (persistence, bioaccumulation, toxicity, and long-range environmental transport), which can distinguish POPs from a multitude of other organic chemicals. All 12 prioritized POPs or their breakdown products rank high to extreme on measurements of these parameters. Reproductive, developmental, behavioral, neurological, endocrine, and immune adverse health effects on people have been linked to POPs. POP pollution has touched every region in the world. Much attention is given to POP contamination problems, and strong action has been taken by most developed and developing nations.

Among organic pollutants, polycyclic aromatic hydrocarbons (PAHs) are the most ubiquitous and constitute a major group of marine environmental contaminants. PAH show a marked hydrophobic character, resistance to biodegradation (Neff, 1979) and adverse effects on health (carcinogenic and/or mutagenic activity) (Singh et al., 1998) and ecosystem (Long et al., 1995). The carcinogenic properties of some compounds, coupled with the stability of PAHs during their atmospheric and aquatic transport, and their widespread occurrence have, in recent years, generated interest in studying their sources, distribution, transport mechanisms, environmental impact and fate (Bouloubassi & Saliot, 1993). Moreover, PAHs are quickly adsorbed onto suspended particulate. These findings prompted US EPA to include 16 PAHs within a priority pollutant list. Therefore, guidelines were proposed in order to assist the management of polluted sediments (Swarz, 1999). PAHs of coastal sediments are due to both anthropogenic and natural sources (NRC, 1985). Among anthropogenic factors, petrogenic and pyrolytic sources are the most important. Although somewhat controversial, the aromatisation of cyclic compounds could be a further source of PAHs (LaFlamme & Hites, 1978). Perylene is a good example of a PAH substance of biological origin found in both marine and freshwater sediments (LaFlamme & Hites, 1978).

Again, PAHs from pyrolytic and petrogenic sources exhibit different chemical behaviour and distribution in marine sediments. In particular, PAHs from pyrolytic processes are more strongly associated with sediments and much more resistant to microbial degradation than PAHs of petrogenic origin (Gustafsson et al., 1997).

The Bizerte lagoon is located in the northernmost part of Tunisia, between latitudes N°37 08′ and N°37 16′ and consists of a depression having a surface area of about 150 km2 and maximum depth of 12 m. This lagoon is known for its geostrategic position since it links the Mediterranean Sea to an internal lake (Ichkeul), which is classified as a national park and world heritage.

Many decades ago, the lagoon is known as a fishery and aquaculture park related to the presence of three mytiliculture sectors especially Menzel Jemil park which is the most productive site. Many years ago, the lagoon's banks were subjected to both urbanization and industrialization. Manufacturing facilities such as iron and steel complex, the naval construction and tyre production, urban wastewater and open-dumping-type municipal or industrial solid waste landfills scattered around the lagoon could have led to ecological perturbation of this fragile environment.

Two previous preliminary studies showed that superficial coastal sediments from Bizerte lagoon were moderately contaminated by PCB (Derouiche et al., 2004) and PAH (Trabelsi & Driss, 2005). In order to perform the current contamination status in the study area data compilation and reporting together with data for POC contamination are the final steps for a comprehensive survey of Bizerte lagoon for the levels of organic contaminants, and represent an attempt to understand the effects of the agricultural and industrial chemical pollution. As a result, the following interpretation and discussion will be focused on the 15 organochlorine pesticides, 20 PCBs and 20 PAHs in sediment. Additionally, data for combination of multivariate approaches were applied in order to identify possible input sources.

2. Materials and methods

2.1 Sampling

Sediment samples from Bizerte lagoon were collected in December 2001 using a Van Veen grap. The first top centimetre (500g) was placed in an aluminium container and frozen on dry ice, then transferred to the laboratory with no exposure to light. Fifteen sampling stations in coastal and central areas in the lagoon are shown in Fig.1.

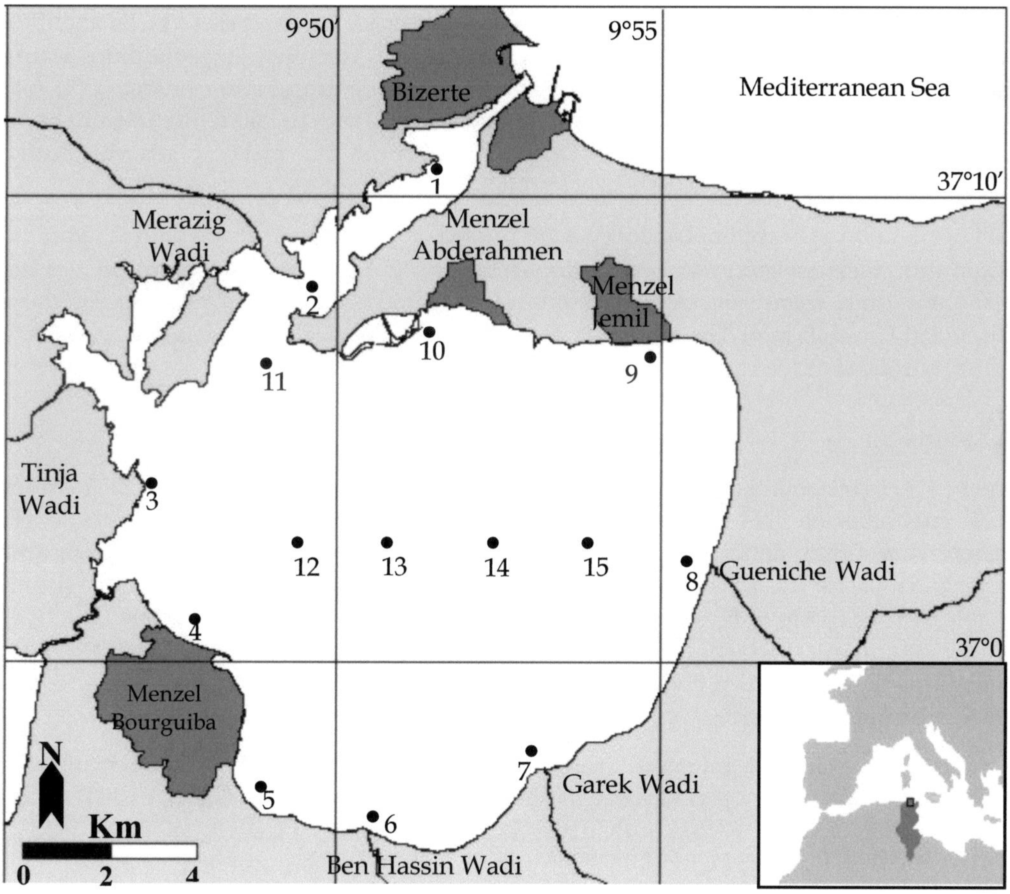

Fig. 1. Study area and sampling stations

2.2 Grain size and TOC determinations

The fraction of the sediment having a grain size <63 μm (silt+clay=pelite) was determined gravimetrically after wet sieving. A representative sub-sample was taken for organic carbon (TOC) determination, according to the revised Walkly–Black titration method conducted in accordance with clause 3 of BS 1377: Part 3 (BSI, 1990). Briefly, 2g dried sediments were treated with 10mL of 1.000N potassium dichromate followed by the rapid addition of 20 mL of concentrated H_2SO_4 containing 0.5g silver sulphate, to precipitate chloride ions. Samples were allowed to cool uniformly to room temperature for 30min (at 20°C), then the mixture was diluted by 200mL of double-distilled water, and 10mL of phosphoric acid was added. Finally, the extra dichromate was back titrated with iron (II) sulphate solution using barium diphenylamine sulphonate as an indicator.

2.3 Instrumental

Quantitative analyses of PAHs, PCBs and OCPs in sediments were accomplished by an Agilent 6890 A gas chromatography (GC). The column used for analysis of PCBs and OCPs was fused silica capillary PTE-5 (30 mx 0.32 μm thickness). The operating conditions were as follows: Injector temperature 250°C; detector 300°C; oven temperature: initial 50°C for 2 min, programmed to 160°C at 5°C/min, followed at 2°C/min to 260°C for 10 min; carrier gas: helium at a flow rate (constant flow) of 1.5 mL/min; detector make-up gas was nitrogen at a flow rate of 60 mL/min; sample injection volume 2μl; injection mode: splitless for 1 min.

PAHs separation was accomplished by a HP1 fused silica column (30m length 0.32mm i.d., 0.25μm film thickness) and flame ionisation detector (FID). The oven temperature was held at 50°C for 1min, then programmed at 20°C/min to 150°C then at 8°C/min to 280°C and held at 280°C for 15 min. The injector was maintained at 280°C and the detector at 300°C. Helium was used as the carrier gas.

2.4 Standards

Twenty PAH compounds were analysed in this study, including sixteen US EPA priority PAHs (naphthalene (Naph), acenaphthylene (Act), acenaphthene (Ace), fluorene (Fl), phenanthrene (Phe), anthracene (An), fluoranthene (Ft), pyrene (Py), benzo[a]anthracene (B[a]an), chrysene (Chy), benzo[k]fluoranthene (B[k]ft), benzo[b]fluoranthene (B[b]ft), benzo[a]pyrene (B[a]py), indeno [1,2,3-cd]pyrene (Ind), dibenzo[a,h]anthracene (D[a,h] an) and benzo[ghi] perylene (B[ghi]pe) (99%, Supelco), perylene (99%) and three alkylatyed PAHs [2-methyl-anthracene (99%), 9-methyl-anthracene (97%) and 1-methyl-phenanthrene) (99%)] from Jansen Chemica.

A mixed PCB standard reference material (SRM 1493) from the National Institute of Standard and Technology (USA) was used for and containing PCB8, PCB18, PCB28, PCB44, PCB52, PCB66, PCB77, PCB101, PCB105, PCB118, PCB126, PCB128, PCB138, PCB153, PCB170, PCB180, PCB187, PCB195, PCB206 and PCB209 in IUPAC number.

The standard reference material SRM 2261, obtained from National Institute of standards and Technology, was used to quantify the 15 examinated organochlorine pesticides: HCB, lindane, Heptachlor, aldrine, heptachlorepoxyde, o,p′-DDE, cis-chlordane, trans-nonachlor, dieldrine, p,p′-DDE, o,p′-DDD, p,p′-DDD, o,p′-DDT, p,p′-DDT and mirex.

2.5 Analytical procedure

The analytical procedure used for PAHs follows the modified method described by Kelly et al. (2000). Briefly, 20g of dried sediments were extracted using alkaline saponification with KOH/MeOH (2N, 100mL). Digests were filtered and extracted with n-hexane, and dried over previously roasted Na_2SO_4. After sulphur removal by activated copper, the solution was reduced to 2mL and dropped into a column (i.d. = 10mm), slurry packed with 2g of activated silica gel 60, then eluted with a sequence of 15mL hexane (fraction 1-discarded) and 30mL of hexane: dichloromethane (9:1; fraction 2) which contained the PAHs. The eluate was concentrated to 0.5mL in a micro-Kuderna-Danish evaporator under a gentle stream of nitrogen.

PAHs were identified and quantified by comparison with known standards injected under the same conditions. Certified standard reference marine sediment (EC-7, National Water Research Institute, Canada) was used in the evaluation of the analytical method. The mean recovery of certified PAHs in the extract was 86%. The detection limit of each PAH has an average of about 0.03 ngg-1 for the sediment samples.

The analytical procedure of OCPs and PCBs in sediments is a modification of the method described by Montone et al. (2001). Briefly, 30 g of dried sediment were extracted with 300 mL of n-hexane in a Soxhlet apparatus for 12 h. Activated copper treatment and sulphuric acid clean-up procedures were employed to remove elemental sulphur and other interfering materials. The sample extracts were further purified in a column (i.d.= 10 mm), slurry packed with 5 g of florisil and then eluted with a sequence of 40 mL hexane (fraction contained OCPs) and 40 mL of hexane: dichloromethane (9:1; fraction 2) which contained the PCBs. Each fraction was concentrated to 1 mL in a micro-Kuderna-Danish evaporator under a gentle stream of nitrogen.

The whole analytical procedure was validated by analysing EC-3 sediment reference materials from National Water Research Institute (Canada). The recoveries of studied PCBs in the extract using the same methodology were>90%. The identification of compounds was deduced from their retention times and quantification was based on peak area measurement as well comparison with responses of the mixed PCBs standard reference material (SRM 1493).

3. Results and discussion

3.1 Sediment characteristics

Different opinions exist regarding the relation between POPs concentration and characteristics of sediments. Several authors have observed that organochlorines and PAH in the sediments are mainly associated with the organic matter (Neff, 1979; Knezovich et al., 1987; Doong et al., 2002), and their total content depends on grain size distribution (Law & Andrulewicz, 1983; Doong et al., 2002).

TOC contents ranged from 0.23% to 2.95% in the study area (Table1). Organic matter contents of sediments appear especially high (> 3% of dry weight) in the central zone of the lagoon and are relatively low in the peripheral zones (<1%). Sediments from stations 4 and 5 which are adjacent to area of Menzel Bourguiba city (about 100000 habitants), known by its big urban and industrial activities, seem to be especially rich in organic matter (>1.5%). This also seems to be true for the Bizerte and neighbouring cities. This finding could be explained by the massive discharge (before 1997) of raw municipal and/or industrial wastewater before implantation of wastewater plants in these cities. The mouths of Guenniche and Garek wadis are marked by a low deposit of organic matter (<0.5%).

The pelite (fraction of the sediment having a grain size <63 μm (silt+clay)) is given in Table1. The samples from Bizerte lagoon show large variations in sediments type, from very coarse (5 % are < 63μm at station 1) to very fine grained (80 % are < 63μm in the central zone of the lagoon). Low pelite content (11%) was also found in sandy sediments from the eastern part (station 9). In mouths of Guenniche, Ben Hassin and Garek wadis, the sediments were less fine-grained with pelite varying from 21 to 37 % of dry sediment.

The major part of Bizerte lagoon basin is a depocenter of fine grey mud generally saturated with sodium ions (sodic clays) having a high cation exchange capacity exceeding 10 meq/100g dry sediment (Hamdi et al., 2002). However, a sandy environment, mainly in the southern part where coarse sediments are mixed with shell debris, characterizes the near-shoreline contour.

The kinetic behavior of hydrophobic organic pollutants is much influenced by organic carbon contents in sediments and soils (Karickhoff et al, 1979).

Our study shows that the concentrations of ∑DDT and ∑POC, in sediments from Bizerte lagoon are correlated well with TOC contents ($r = 0.553$, $p < 0.05$), which is possibly due to the less lipophilic and more volatile nature of HCHs relative to the DDTs.

Correlation analysis shows negative correlations between total PAHs contents/TOC contents and the pelite/PAH levels. These results indicate that the observed distribution of PAHs was not governed by sedimentary characteristics such as TOC contents and the fraction of pelite (grain size less than 63μm), but, it might be due to the localised source inputs. However, positive and significant usual correlations (R_2=0.70) were observed between the sediment contents of fine fraction and those of TOC contents (Johnson, 1986). Indeed this association between fine matter and the organic particles is especially due to a common phenomenon in Mediterranean lagoons called flocculation (Aloisi et al., 1975).

3.2 PAH Composition source identification and ecotoxicological concerns

The distribution of various PAH in the sediments from 15 sampling stations reveals a wide range of fluctuations, as delineated in Table 1.Twenty individual PAHs were determined which comprising two to six ring, with three alkyl-substituted homologues. Total PAHs concentrations ranged from 20 at station 15 to 449 ngg^{-1} dry weight at station 5 with a mean average of 183 ngg-1 dry wt. The lower concentrations were detected in the middle part of the lagoon and the eastern zone which is characterized by an agricultural activity and where several wadis flow directly into the lagoon. The highest contents are found at stations 1, 2, 4, 5, 6 and 11 where an intensive urban and/or industrial activities were carried out.

Certain diagnostic ratios have been widely used in order to identify and quantify the contribution of each sources of PAH contamination. The reported approaches should be treated with caution.

	Stations														
PAH	1	2	3	4	5	6	7	8	9	10	11	12	13	14	15
Na ph	5.42	3	3.26	3.44	10.02	nd	3.3	9.54	nd	12.82	6.69	2.14	nd	Nd	nd
Ace	0.28	0.77	0.89	0.88	0.24	nd	3.29	4.65	0.27	0.44	0.26	0.42	0.93	2.57	nd
Act	nd	nd	nd	nd	1,45	nd	nd	nd	nd	nd	nd	nd	nd	Nd	nd
Fl	nd	nd	nd	nd	7,8	nd	nd	nd	nd	nd	nd	nd	nd	Nd	nd
2-me-An	0.08	0.12	0.18	0.5	0.1	0.32	0.15	0.07	nd	0.05	0.12	nd	nd	Nd	nd
1-mé-Phe	0.73	1.87	3	nd	1.91	1 .98	nd	0.36	0.36	nd	nd	nd	nd	Nd	nd
9-me-An	0.08	0.09	0.08	1.4	0.06	nd	nd	Nd	nd	nd	0.7	0.07	nd	Nd	nd
Phe	30.85	19.97	26.61	27.1	54.64	nd	nd	Nd	13.93	27.19	13.08	29.96	22.71	30.81	7.99
An	2.8	nd	7.72	13.72	30.68	45.48	16.05	20.63	3.49	6.4	1.75	nd	nd	2.9	nd
Ft	44.2	96.48	nd	70.09	45.86	49.77	20.16	36.78	11.23	17.34	26.37	22.32	18.96	19.08	8.66
Py	91.99	64.37	21.7	33.7	176.06	nd	nd	Nd	14.1	9.51	26.89	19.52	17.67	Nd	nd
B(a) an	38.47	65.65	nd	nd	33.04	nd	nd	Nd	nd	nd	nd	9.78	16.53	7.96	nd
Chy	27.8	52.08	nd	140.94	nd	34.68	nd	Nd	72.78	nd	106.49	11.52	10.44	9.22	nd
B(b) ft	32.33	2.43	nd	8.72	nd	31.74	nd	Nd	nd	34,8	48.77	11.2	nd	Nd	3.76
B(k) Ft	nd	nd	nd	60.73	nd	nd	nd	Nd	nd	nd	4.16	nd	nd	Nd	nd
B(a) py	20.18	51.6	39.56	36.67	35.54	nd	13.03	Nd	nd	18.96	nd	20.86	nd	Nd	nd
Per	0.05	0.05	nd	0.08	nd	0.68	nd	0.06	nd	nd	0.28	0.09	0.07	0.06	nd
Ind	nd	36.75	nd	nd	42.31	nd	nd	Nd	nd	nd	nd	nd	nd	Nd	nd
D(a,h) an	nd	nd	nd	nd	3.11	nd	nd	Nd	nd	13.61	nd	nd	nd	Nd	nd
B(ghi) pe	nd	nd	nd	nd	6.44	nd	nd	Nd	nd	nd	nd	nd	nd	Nd	nd
Tot PAH	295.26	421.23	103	397.97	449.26	164.65	55.98	72.09	116.16	141.12	235.56	127.88	87.31	72.6	20.41
OC (%)	0.3	1.85	1.91	1.87	1.87	0.87	0.23	0.36	0.95	0.9	0.82	1.8	2.6	3.2	2.1
Pelite (g)	5.03	42.04	64.23	88.51	59.98	22.75	37.58	21.47	11.29	23.13	51.56	86.24	91.76	96.41	89.57

Table 1. Levels of PAHs (ngg-1 dry wt), percentage of organic carbon (%OC) and pelite fraction (g) in surface sediments from 15 sampling stations

In contrast to pyrogenic sources, petrogenic sources are characterized by Phe/An ratio>10, Ft/Py <1 and B(a)An/Chy<1 (Budzinski et al., 1997). On the other hand, some PAHs, such as perylene come from diagenesis processes of biogenic precursors (Venkatesan, 1988).

Recently, alkylated PAHs also, have been proved useful in petroleum-related fingerprinting by comparing their relative individual magnitudes in sediments (Page et al., 1999).

Budzinski et al., (1997) had given attention to distribution of low and high molecular weight PAH (LPAH and HPAHs, respectively) as a reliable tool for discriminating the petrogenic/pyrolytic origin of PAHs. The higher the LPAHs/HPAHs ratio, the higher the prevalence of petrogenesis on pyrolytic origin of PAHs.

Concentrations of PAHs are mostly dominated by high molecular weight compound (PAHs with 4 to 6 rings). The LPAH/HPAHs ratios are very low, from 0.06 to 1 and the only congener that is nearly always present is phenanthrene. This means that probably high temperature combustion processes were the predominant source of PAHs contamination (Canton and Grimalt, 1992). This information is confirmed by the low concentrations of the alkyl homologues (methyl-anthracene/anthracene and methyl-phenanthrene/phenanthrene). Indeed, close similarities between the relative distribution of methyl-anthracene/anthracene and methyl-phenanthrene/phenanthrene ratios for sample sediments analyzed in the study area were observed but less than unit values of alkyl to parent anthracene and phenanthrene were detected for all sample sediments.

Additionally, the Phe/An and Ft/Py ratios can be used to assess the relative importance of the sources and the effect of diagenetic changes (Soclo et al., 2000). Phe/An ratios span from 3.45 to 11 and Ft/Py ratios range from 0.26 to 2. For instance Phe/An>10 and Ft/Py<1, which are characteristics of PAH petrogenic contamination, were found only at station 1. Furthermore, chrysene and B(a)an derive from pyrolytic sources in ratio that should be lower than 1 (Soclo et al., 2000). Most values in our samples were closed to 1 or slightly higher. This may be due to some influence from petrogenic contamination or to the degradation of the less stable B(a)An.

Perylene, which may be of natural origin, was determined at very lower concentrations (Table1). Early studies reported diagenetic formation of perylene from terrestrial precursors in anoxic conditions (Aizenshtat, 1973). However, perylene is also produced during pyrolytic processes (Venkatesan, 1988). The low relative abundance of perylene in the study area argues for a dominant pyrolytic origin (Venkatesan, 1988).

Distribution of PAHs in Bizerte Lagoon sediment reveals general and moderate contamination. Based on the investigation of approaches reported above our results reveal that PAH contamination could be related to urban and or industrial activities. The data found indicate that sediments for stations 1, 2 and 11 were characterized by pyrogenic PAH contamination when examining the LPAHs/HPAHs, methylphenanthrene/phenanthrene, methylanthracene/anthracene ratios. However, Phe/An and Ft/Py ratios reveal PAH petrogenic contamination only at station 1. PAH inputs in this area are most likely related to the intensive traffic of boats and industrial activities. The later station is localised in an area where there are oil refinery and used oil recycling activities which probably constitute the main sources of contamination by petrogenic PAHs. Stations 2 and 11 are near a base of the Tunisian Military Marine and a solid waste landfill. Stations 4, 5 and 6 are located near Menzel Bourguiba city, where intensive industrial activity is carried out, including the metallurgic industry, naval construction and tyre production. In addition, there are several local wastewater discharges in this area. Additionally, PAH concentrations for all stations were also plotted using a ternary diagram to investigate combinations of PAHs that may have similar sources and modes of inputs. The plot reveals two notable groups of stations. The first includes stations 1, 2 and 11 which are located in the channel of navigation of boats. In the second group, we distinguish stations 4, 5 and 6, which are closed to the urban and intensive productive industrial zone (city of Menzel Bourguiba).

The middle part (12, 13, 14 and 15) of the lagoon shows a clear tendency of PAH concentration to decrease from west to east part. The decrease of PAH levels is logically in agreement with the increasing distance from urbanised and/or industrial area.

To assess potential environmental impacts of PAHs in sediments, the concentrations of PAHs in sediments can be compared with certain threshold values proposed by the US National Oceanic and Atmospheric Administration, e.g. the effects-range low (ER-L) (which is a sediment quality guideline below which the chemical concentration in the sediments would be expected to rarely have adverse biological effects) and the effect-range medium (ER-M) (a level above which adverse biological effects occur frequently).

The concentrations found for all stations ranged from 20 to 449 ngg-1 dry weight which are below the ER-L (4000 ngg-1) and the ER-M (45000 ngg-1). This range represents a relatively clean environment.

3.3 Contamination status of POPs

3.3.1 PCB composition and sources identification and ecotoxicological concerns

The distribution of various PCB in the sediments from 15 sampling stations reveals that the ΣPCB ranged from 0.89 to 6.63 ngg^{-1} (Table 2). Fig.2 shows the geographic distribution of the ΣPCB in the whole study area, the contents of total PCB in sediment vary from a site to another.

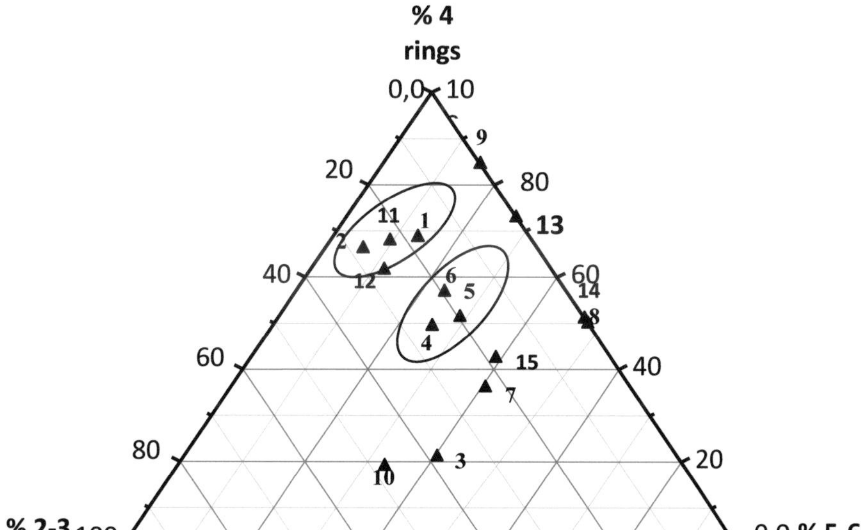

Fig. 2. ternary diagram of the distribution of low, mid and high molecular weight PAHs

This result reflects geographic distribution of PCB level of contamination. For the coastal sites, the highest PCB concentrations ranged from 4.39 to 6.63 ng g^{-1} were detected at northern part covered sites 1, 2, 10, and 11 and at southwest zone (site 5). Sediments collected from the other coastal sites are slightly contaminated (∑PCB ranged from 0.89 to 2.58 ng g-1). Relatively high PCB concentrations were also found in samples collected from sites 12, 13, 14, and 15. These stations are located in the central zone of the lagoon and are the deepest.

	Stations														
PCB	1	2	3	4	5	6	7	8	9	10	11	12	13	14	15
PCB8	0.23	0.31	0.05	0.46	0.18	0.33	0.40	0.34	0.36	0.76	0.28	0.69	0.86	1.38	0.96
PCB18	nd	nd	nd	nd	nd	nd	nd	nd	0.48	0.31	0.19	0.37	0.31	0.80	0.45
PCB28	0.15	0.22	nd	0.24	nd	nd	0.23	0.19	0.22	0.35	0.20	0.32	0.31	0.52	0.32
PCB 52	0.11	0.44	nd	0.15	nd	0.07	0.09	0.07	0.08	0.32	0.22	0.36	0.13	0.10	0.08
PCB 44	nd	0.21	nd	nd	nd	0.06	nd	nd	0.10	0.11	nd	0.12	0.10	0.10	0.06
PCB 66	nd	0.10	0.09	nd	0.03	nd	0.06	nd	0.11	0.17	0.36	0.20	0.12	0.15	0.07
PCB 77	nd	nd	nd	nd	nd	nd	nd	nd	0.41	0.29	0.44	0.99	0.18	0.26	0.09
PCB 101	0.32	0.64	nd	0.15	0.19	0.06	0.07	nd	0.07	0.25	0.31	0.23	0.37	0.11	nd
PCB 118	nd	0.54	0.26	0.09	0.09	nd	nd	nd	0.17	0.17	0.38	0.27	0.16	0.12	0.07
PCB 105	0.15	0.23	0.15	0.08	0.07	0.02	nd	nd	nd	0.06	0.15	0.08	nd	0.08	nd
PCB 153	0.80	0.73	0.09	0.40	1.09	0.23	0.10	0.08	0.10	0.51	0.91	0.78	0.62	0.37	0.34
PCB 138	0.95	0.85	nd	0.38	1.01	0.19	0.08	0.07	nd	0.40	0.72	0.58	0.40	0.32	0.25
PCB 128	0.17	0.29	nd	nd	0.18	0.03	nd	nd	nd	0.04	0.08	0.04	0.04	0.07	0.08
PCB 187	0.22	0.15	0.11	0.12	0.36	0.07	nd	nd	0.06	0.22	0.55	0.40	0.27	0.22	0.19
PCB 180	0.70	0.34	0.19	0.23	0.94	0.15	nd	0.05	0.25	0.41	1.18	0.76	0.55	0.27	0.23
PCB 170	0.50	0.20	0.12	0.15	0.57	0.07	nd	nd	0.06	0.18	0.44	0.27	0.16	0.14	0.12
PCB 195	0.09	0.09	nd	0.10	nd	nd	0.09	0.08	nd	0.05	0.09	0.04	0.03	0.02	0.02
PCB 206	nd	nd	nd	nd	nd	nd	nd	nd	nd	0.01	0.03	0.03	0.01	nd	nd
PCB 209	nd	nd	nd	0.04	0.07	nd	nd	nd	nd	0.08	0.10	0.10	0.09	0.09	0.07
∑PCB	4.39	5.34	1.51	2.58	4.77	1.29	1.12	0.89	2.48	4.68	6.63	6.62	4.69	5.13	3.40

Table 2. Concentrations (ngg^{-1}, dry weight) of PCBs in surface sediments from 15 sampling stations

Our sediment investigation revealed the general prevalence of lower and mid molecular weight PCBs. The PCB congeners pattern was dominated by bi-PCBs (100%), Tri-PCB (PCB28; 80%) and tetra-PCB (PCB52; 86%). Mid-PCBs also dominated the PCBs congeners pattern. PCB congeners proportions consisted of the following distribution: PCB153 (100%), PCB180 (93%) and PCB (187, 170, 138) accounting 86% of total PCBs composition. However, relatively moderate percentage of the high chlorinated congeners, such as octa-PCBs (73%), were found in this study. Sediment collected at locations 4, 5, 6, 14 and 15 contained great proportions of di-, tri-, and tetra-PCB, whereas those from locations 1, 2, 11, 10, 9,8 and 7 had higher proportions of penta-, hexa-, and hepta-PCBs.

A number of previous studies have showed the production of PCBs during the steel manufacturing processes (Alcock et al., 1999; Buekens et al., 2001). This is due to the presence of PCBs in fly ash generated from burning coal during the iron ore sintering process (Biterna & Voutsa, 2005). The later sites (1, 2, 11, 10, 9, 8 and 7) are close to some industries established on the circumference of the lagoon (an iron and steel plant, a cement factory, and a refinery).

The human population around the lagoon is estimated at 163,000 inhabitants (census of 2004) of which approximately 70% are concentrated in Bizerte town. The other main important towns bordering the lagoon are Menzel Bourguiba (which has a naval port and a metal factory), Menzel Abderrahmen and Menzel Jemil. The sources of untreated industrial and domestic wastewater of these cities, which are indicative of the near-source emissions of

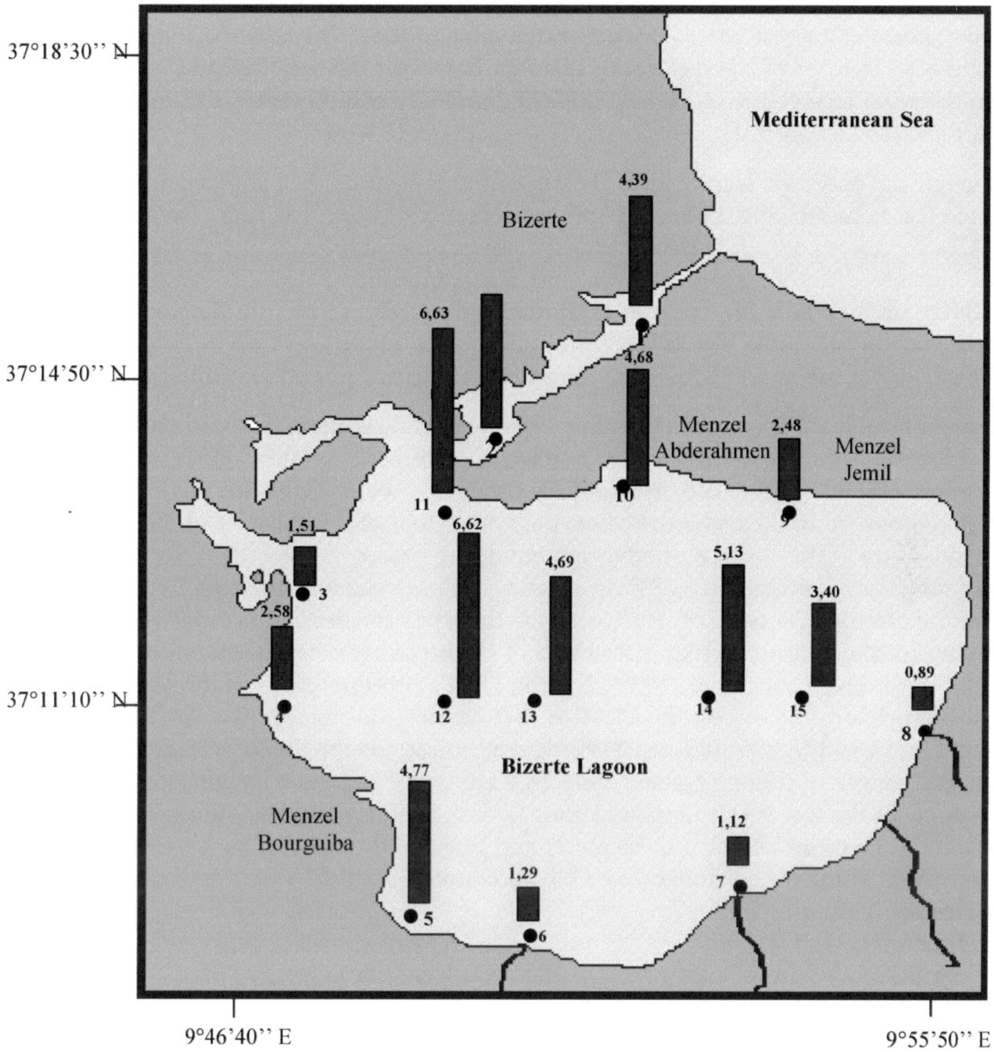

Fig. 3. Spatial distribution of ∑PCBs in surface sediment from Bizerte lagoon

PCBs from industrial wastewater and domestic sewage, are also, the main contributors. PCBs are an industrial product; there are no known natural sources. Atmospheric depositions, runoff from the land, and food chain transport (Morrison et al., 2002; Totten et al., 2006; Davis et al., 2007) have been regarded as the major sources of PCBs in aquatic environments. Urban runoff from local watersheds is a particularly significant pathway for PCB entry into the lagoon. Since PCBs are somewhat volatile and tend to enter the atmosphere, atmospheric transport and deposition can be important processes, such as exchange between the water and the atmosphere, and between the soil and the atmosphere.

A fraction of the PCBs lost by this pathway may return to the water and land surface via deposition in the watershed and subsequent runoff. Besides, PCBs have been shown to bioconcentrate significantly in aquatic organisms (Morrison et al., 2002). Four open-dumping-type municipal or industrial solid waste landfills are scattered around the lagoon; the first and the biggest one is located on the edge of the canal near the cement works. The second one, in front of Menzel Bourguiba city however, the smallest ones and the farthest from the coast receive the garbage of Menzel Jemil city could escort to an important source of PCBs in the lagoon.

Although the observed trends in PCB congener composition was similar in some sediment sample (i.e. between sites 12 and 13 and between sites 14 and 15). The distribution patterns of PCBs congeners are, in general, different among the sediments of this study area, which may indicate different input sources and the establishment of correlation between the congener profiles and the sources is difficult, especially when the distance between the source and the sampling site is large. In addition the usage of PCBs in Tunisia is not well established, but the use of PCBs in transformers, electrical, and other industries is common.

Sediment-bound PCBs can affect benthic organisms. To evaluate the ecotoxicological aspect of sediment contamination, some published sediment quality guidelines and toxic equivalent (WHO-TEQ) of dioxin-like PCBs congeners were applied in this study. Although the guidelines are limited in some cases, they provide useful indicators of the effects of PCB contamination in the absence of environmental assessment criteria for PCBs in Tunisia. The effect range low value (ERL, 22.7 ng g-1 dry weight) suggests that PCBs can exert toxic biological effects on aquatic organisms, while the effect range median value (ERM, 180 ng g-1 dry weight) indicates the high possibility of PCBs posing detrimental biological effects on aquatic organisms (Long et al., 1995). The total PCBs concentrations of the samples collected for this study do not exceed the ERM or ERL values. As regards the results obtained for dioxin-like PCBs (77, 126, 105 and 118) the concentrations for the PCB77, 105 and 118 vary from<0.02 ngg-1 to 0.99 ngg-1 and only PCB126 is not detected for all sediment samples. Moreover, in the lower contaminated sites (7 and 8) all the studied dioxin-like PCBs were not found. The ratios between these levels and those of the total PCBs were dissimilar in all the locations studied; the dioxin-like PCBs accounted for 0-23% of the total PCBs in these samples.

3.3.2 OCPs composition and sources identification and ecotoxicological concerns

In case of OCPs, only 10 coastal sampling points were considered because concentrations for the most analysed pesticides were below the detection limit (<0.010) (Table 3). The obtained results showed that HCB and DDT isomers (DDT+DDE+DDD) were the predominately detected compounds in most stations with concentrations ranged between 0.02 to 0.12 ngg^{-1} and 0.21 to 3.74 ngg^{-1}, respectively. Heptachlor was the second highest with concentration ranged between 0.03-0.3 ngg^{-1} whereas, lindane and mirex were detected only in one station. The other pesticides were not detected in any sample.

The most contaminated stations were 2 and 3 with total pesticides concentration 3.78 ngg-1 and 3.47 ngg^{-1}, respectively. Sites 6 (0.18 ngg^{-1}) and 7 (0.23 17 ngg^{-1}) were the less contaminated stations. The other sites were moderately polluted with total pesticides concentration ranged between 0.56 ngg^{-1} in station 10 and 1.83 ngg^{-1} in station 5.

The composition of organochlorines and their metabolities can provide some information for a better understanding of the origin and transport of these contaminants in the environment. Microbial degradation of DDT, DDD and DDE is generally slow, resulting in environmental persistence of these compounds and DDT may degrade to DDD with a half-life of a few days under certain conditions (Garrison et al., 2000). DDT can be biodegraded to DDE under aerobic condition and to DDD under anaerobic condition (Kalantzi et al., 2001). Comparing the concentrations of ∑DDT and its metabolites, it can be inferred whether DDTs input are recent or not (Phuong et al., 1998). The ratio of (DDE + DDD)/∑DDT > 0.5 can be thought to be subjected to a long-term weathering (Hites & Day, 1992; Hong et al., 1999). In most present samples, DDE and DDD occupied the predominant percentage. Here the mean ratio of (DDE + DDD)/∑DDT in the sediments from Bizerte lagoon was 0.85, which indicated that the degradation occurred significantly. However, the DDT (including o,p-' and p,p'-) in the sediment of site 3 occupied about 50%. The high levels of ∑DDT and high percentage of DDT might due to intensification of the agricultural activities in areas surrounding the lagoon. In general, potential sources of OCPs pollution in the study area can be: the use of these pesticides in the past in agricultural area surrounding the lagoon, domestic sewage and atmospheric transport.

	Stations									
OCPs	1	2	3	4	5	6	7	8	9	10
HCB	0.03	0.02	0.10	0.09	0.06	0.03	0.02	0.09	0.12	0.12
Lindane	nd	nd	nd	nd	nd	0.09	nd	nd	nd	nd
Heptachlor	0.27	0.03	0.06	nd	nd	0.06	nd	0.36	0.14	0.06
o,p'-DDT	0.05	0.48	0.29	nd	0.32	nd	nd	nd	0.03	nd
p,p'-DDT	nd	0.85	1.33	nd	0.17	nd	nd	nd	nd	nd
o,p'-DDE	nd	0.06	0.15	0.14	0.13	nd	nd	0.13	0.22	nd
p,p'-DDE	0.14	0.18	0.43	0.37	0.21	nd	nd	nd	0.16	0.13
o,p'-DDD	0.33	0.99	0.48	0.38	0.40	nd	0.21	0.49	0.48	0.35
p,p'-DDD	0.09	1.17	0.60	0.08	0.51	nd	nd	nd	0.02	nd
∑DDT	0.61	3.73	3.28	0.97	1.74	0	0.21	0.62	0.91	0.48
∑OCPs	0.91	3.78	3.44	1.06	1.8	0.18	0.23	1.07	1.17	0.66

Table 3. Concentrations (ngg^{-1}, dry weight) of OCPs in surface sediments from 10 sampling stations

4. Conclusion

In Bizerte lagoon, Tunisian area economically very important, sediments appear moderately polluted. The contamination level appears rather low compared to those found in other ecosystems.

On the basis of this study, we can conclude that the lagoon is subject to intensive industrial activities. There are four main zones of anthropogenic influence: In zone 1 are situated oil refineries, food and ceramic industries. In zone 2 are located cements, treatment of metals (copercraft, asbestos) and sprinkling beverage factories. In zone 3 are ceramic and metallurgy activities. In zone 4 are present metallurgy activities (Fe, Zn, Cd, Sn, Hg), naval constructions and tire productions.

Because Tunisia has a long Mediterranean coast of crucial economical interest, it is obvious that a larger spatio-temporal POP monitoring has to be planned, including harbour areas.

This is the condition to have an evaluation of the actual organic pollutants contamination of the Tunisian marine environment.

5. References

Aizenshtat, Z. (1973). Perylene and its geochemical significance. *Geochim. Cosmochim. Acta.*, vol.37 pp.559-567

Alcock, RE.; Gemmill, R. & Jones, KC. (1999). Improvements to the UK PCDD/F and PCB atmospheric emission inventory following an emissions measurement programme. *Chemosphere,* vol. 38 pp.759–770

Aloisi, JC. ;, Monaco A. & Pauc, H. (1975). Mécanismes de la formation des prodeltas dans le Golf du lion. Exemple l'embouchure de l'Aude (Languedoc). *Bull. Inst. Géol. Bassin Aquitaine,* vol.18 pp.3-12

Bayer, M. (1959). *Gas chromatography,* Springer-Verlag, Berlin (proved the 450 years old work of Brunschwig, a Strassburg surgeon)

Birnbaum, LS. (1994). Endocrine effects of prenatal exposure to PCBs, dioxins, and other xenobiotics: implications for policy and future research. *Environ. Health Perspect,* vol.102 pp.676–9

Biterna, M. & Voutsa, D. (2005). Polychlorinated biphenyls in ambient air of NW Greece and in particulate emissions. *Environmental International,* vol. 31 pp. 671–677

Bouloubassi, J. & Saliot, A. (1993). Investigation of anthropogenic and natural organic inputs in estuarine sediments using hydrocarbon markers (NAH, LAB, PAH). *Oceanol Acta,* vol. 16 pp.145–61

BSI. BS 1377: 1990 British Standard Methods of Tests for Soils for Civil Engineering Purposes, Part 3 Chemical and Electrochemical Tests, British Standard Institution, London. 1990

Budzinski, H.; Jones, I.; Bellocq, J.; Pierad, C. & Garrigues, P. (1997). Evaluation of sediment contamination by polycyclic aromatic hydrocarbons in the Gironde estuary. *Mar. Chem.*, vol.58 pp. 85– 97

Buekens, A.; Stieglitz, L.; Hell, K.; Huang, H. & Segers, P. (2001). Dioxins from thermal and metallurgical processes: recent studies for the iron and steel industry.*Chemosphere,* vol. 42 pp. 729–735

Davis, JA.; Hetzel, F.; Oram, JJ. & McKeeet, LJ. (2007). Polychlorinated biphenyls (PCBs) in San Francisco Bay. *Environmental Research,* vol. 105 pp. 67–86

Derouiche, A.; Sanda, YG. & Driss, MR. (2004). Polychlorinated Biphenyls in Sediments from Bizerte Lagoon, Tunisia. *Bull. Environ. Contam. Toxicol.*, vol.73 (5) pp. 810–817

Doong, RA.; Peng, CK.; Sun, YC. & Liao, PL. (2002). Composition and distribution of organochlorine pesticide residues in surface sediments from the Wu-Shi River estuary, Taiwan. *Mar. Pollut. Bull.*, vol.45 pp.246–53

Garrison, A.; Nzengung, V.; Avants, J.; Ellington, JJ.; Hones, WJ.; Rennels, D. & Wolfe, NL. (2000). Phytodegradation of p.p-DDT and the enantiomers of o,p′-DDT. *Environ. Sci. Technol.*, vol.34 pp.1663–1670

Gustafsson, O.; Haghseta, F.; Chan, C.; MacFarlane, J. & Gschwend, PM. (1997). Quantification of the diluite sedimentary sootphase: implication for PAH speciation and bioavailability. *Environ. Sci. Technol.*, vol.31 pp. 203– 209

Hamdi, H.; Jedidi, N.; Yoshida, M.; Mosbahi, M. & Ghrabi, A. (2002). Some physico-chemical properties of lake Bizerte sediments. Study on the environment pollution of Mediterranean coastal lagoons in Tunisia. *Initial report*, pp. 49-54

Hansen, JC. (1998). The human health programme under AMAP AMAP human health group Arctic Monitoring and Assessment Program. *Int J Circumpolar Health*, vol.57 pp.280- 91

Hites, RK. & Day, HR. (1992). Unusual persistent of DDT in some western USA soils. *Bull. Environ. Contam. Toxicol.*, vol.48 pp.259–264

Hong, H.; Chen, W.; Xu, L.; Wang, X. & Zhang, L. (1999). Distribution and fate of organochlorine pollutants in the Pearl River Estuary. *Mar. Pollut. Bull.*, 39 vol. pp.376–382

James, AT. & Martin, AJ. (1952). *Biochem. J.*, vol.50 p.679

Kalantzi, QI., Alcock, RE., Joneston, PA., Santillo, D., Stringer, RL., Thomas, GO. & Jones, KC. (2001). The global distribution of PCBs and organochlorine pesticides in butter. *Environ. Sci. Technol.*, vol.35 pp.1013–1018

Karickhoff, SW.; Brown, DS. & Scott, TA. (1979). Sorption of hydrophobic pollutants on natural sediments. *Water Research*, vol. 13 pp.241-248

Kelly, CA.; Law, RJ. & Emerson, HS. (2000). Methods of analysing hydrocarbons and polycyclic aromatic hydrocarbons (PAH) in marine samples. The Centre for Environment, Fisheries and Aquaculture Science (CEFAS) Science Series, Aquatic Environment Protection: *Analytical Methods* No. 12, Lowestoft, England

Knezovich, JP.; Harrison, FL. & Wilhelm, RG. (1987). The bioavailability of sedimentsorbed organic chemicals: a review. *Water Air Soil Pollut.*, vol.32 pp.233–45

LaFlamme, RE. & Hites, RA. (1978). The global distribution of polycyclic aromatic hydrocarbons in recent sediments. *Geochim. Cosmochim. Acta*, vol.42 pp. 289–303

Law, R. & Andrulewicz E. (1983). Hydrocarbons in water, sediment and mussels from the Southern Baltic Sea. *Mar. Pollut. Bull.* vol.14 pp.289–93

Long, ER.; MacDonald, DD.; Smith, SL. & Calder, FD. (1995). Incidence of adverse biological effects within ranges of chemical concentrations in marine and estuary sediments. *Environmental Management*, vol. 19 pp.81–97

Montone, RC.; Taniguchi, S. & Weber RR. (2001). Polychlorinated Biphenyls in Marine Sediments of Admiralty Bay, King George Island, Antarctica. *Mari. Pollut. Bull.*, vol. 42 pp.611-614

Morrison, HA.; Whittle, DM., & Haffner, GD. (2002). A comparison of the transport and fate of polychlorinated biphenyl congeners in three Great Lakes food webs. *Environmental Toxicology and Chemistry*, vol. 21 pp. 683–692

National Research Council (NRC) (1985). Oil in the Sea: Inputs, Fates and Effects. *National Academic Press*, Washington, DC, 1985

Neff, JM. (1979). Polycyclic aromatic hydrocarbons in the aquatic environment sources, fates and biological effects. *London7 Applied Science Publishers*, 262 pp.

Page, DSS.; Boehm, PD.; Douglas, GS.; Bence, EA.; Burns, WA. & Mankiewicz, PJ. (1999). Pyrogenic polycyclic aromatic hydrocarbons in sediments record past human activity: a case study in Prince William sound, Alaska. *Mar. Pollut. Bull.*, vol.38 pp. 247-260

Phuong, PK.; Son, CPN.; Sauvain, JJ. & Tarradellas, J. (1998). Contamination by PCB_s, DDT_s and heavy metals in sediments of Ho Chi Minh City_s canals, Vietnam. *Bull. Environ. Contam. Toxicol.*, vol.60 pp.347–354

Stockholm Convention on Persistent Organic Pollutants. UNEP: persistent organic pollutants 2001. http://www.pops.int/documents/convtext/ convtext_en.pdf

Swarz, R. (1999). Consensus sediment quality guidelines for polycyclic aromatic hydrocarbons mixtures. *Environ. Toxicol. Chem.*, vol.18 pp. 780- 787

Tietz, NW. (1970). *Fundamentals of Clinical Chemistry*, WB. Saunder & Co. Philadelphia, Pa., p.118

Totten, LA.; Panangadan, M.; Eisenreich, SJ.; Cavallo, GJ. & Fikslin, TJ. (2006). Direct and indirect atmospheric deposition of PCBs to the Delaware River Watershed. *Environmental Science Technology*, vol. 40 pp. 2171–2176

Trabelsi, S. & Driss, MR. (2005). Polycyclic aromatic hydrocarbons in superficial coastal sediments from Bizerte Lagoon, Tunisia. *Mar. Pollut. Bull.*, vol.50 pp. 344–359

Venkatesan, MI. (1988). Occurrence and possible sources of perylene in marine sediments-a review. *Mar. Chem.*, vol.25 pp. 1-27